THE DYNAMIC FIRE CHIEF

THE DYNAMIC FIRE CHIEF
PRINCIPLES FOR ORGANIZATIONAL MANAGEMENT

Fire Engineering®
BOOKS & VIDEOS

CRAIG A. HAIGH

Disclaimer
The recommendations, advice, descriptions, and methods in this book are presented solely for educational purposes. The author and publisher assume no liability whatsoever for any loss or damage that results from the use of any of the material in this book. Use of the material in this book is solely at the risk of the user.

Copyright © 2022 by
Fire Engineering Books & Videos
110 S. Hartford Ave., Suite 200
Tulsa, Oklahoma 74120 USA

800.752.9764
+1.918.831.9421
info@fireengineeringbooks.com
www.FireEngineeringBooks.com

Senior Vice President: Eric Schlett
Vice President: Amanda Champion
Operations Manager: Holly Fournier
Sales Manager: Joshua Neal
Managing Editor: Mark Haugh
Production Manager: Tony Quinn
Developmental Editor: Chris Barton
Cover Designer: Brandon Ash
Book Designer: Robert Kern, TIPS Technical Publishing, Inc., Carrboro, NC

Library of Congress Cataloging-in-Publication Data Available on Request

All rights reserved. No part of this book may be reproduced, stored in a retrieval system, or transcribed in any form or by any means, electronic or mechanical, including photocopying and recording, without the prior written permission of the publisher.

Printed in the United States of America

1 2 3 4 5   26 25 24 23 22

To my companion, encourager, champion, and prayer warrior, you have prayed me through and out of more situations than I know. This book is dedicated to my wife, Elizabeth A. Haigh, who taught me years ago to have the courage to rely completely on God, for He will make our path straight. The journey getting to this book is not mine alone but your story as well. My prayer as we continue serving the Lord together—
Find us Faithful!

Ecclesiastes 4:9–12

CONTENTS

Foreword xiii
Acknowledgments xv
Introduction. xvii

1 The Importance of Visioning. 1
    Welcome to My New Reality 2
    Called—But Not Yet Equipped. 3
    Figuring It Out 3
    Where to Start? 4
    Visioning 4
    Learning from Others. 5
    Finding Case Studies. 6
    The Untouchables 7
    My Own Group of Untouchables 7
    Why We Struggle with Vision 9
    Evaluating Your Why 10
    Time Demands 10
    The Jealous Mistress That Is Killing Us 11
    Don't Go It Alone 12
    Gone ~~Fishing~~ Visioning 12
    The Message Matters 13
    Messaging Misconceptions 13
    Avoiding the Underwater Rocks 14
    References. 15
    Questions for Further Research, Thought, and Discussion 16

2 Strategic Planning. 19
    East Dubuque Example. 19
    Learning the Importance of Planning 21
    It Is Not That Difficult 21
    Setting A High Standard 22
    Identified Needs. 22
    Grit 23
    Planning and Execution. 24
    We All Make Plans. 25
    Making Lists. 25
    Eye for the Future 26
    Example: Replacing a Water Main. 26
    SWOT Analysis 27
    Conducting the SWOT Analysis. 29

    The Use of an Outside Facilitator 30
    Knowing When You Need Help 31
    Asking Your Constituents 33
    Writing Goals 33
    Goal Formatting 33
    Plan Time Frame 34
    Professional Look of the Plan 34
    What's Inside 36
    Financial Planning Documents 36
    The Value of Photos 38
    Managing the Plan 38
    Budgeting and Finances 39
    Change Management 39
    Celebrating the Success 40
    Worth the Effort 40
    References 41
    Questions for Further Research, Thought, and Discussion 42

3 Managing Money **45**
    Trusted with a Little . . . Trusted with a Lot 46
    Envelope Financial Management 48
    Understanding Your Role 50
    Strategic Planning's Role in Financial Management 51
    Revenue: Taxes 52
    When Tax Dollars Fall Short 54
    Alternative Revenues 54
    One-Time Revenue 55
    Sustained Revenue 57
    Fines and FD Police Powers 62
    Expanded Services 63
    Budget Tracking and Management 64
    Future Expenditure Planning 66
    Sinking Funds/Escrow Accounts 68
    A Word of Caution 72
    Final Word 73
    References 73
    Questions for Further Research, Thought, and Discussion 75

4 Hiring the Team **79**
    The Volunteer Dilemma 80
    The Good Employee 80
    Seeing the Big Picture 81
    Who's on Your Hiring Team? 82
    Recruitment: Paid Personnel 83
    Recruitment: Volunteer Personnel 84
    Are Volunteers Employees? 85
    The Application 86
    Information on the Application 87

New Candidate Orientation 90
Physical Ability Test 92
Written Testing 94
Interview: Knowing What to Ask. 95
Interview: The Process. 96
Interview Team. 101
Preference Points. 102
Eligibility List 103
Post-Offer Testing. 103
Lasting Legacy. 112
References. 112
Questions for Further Research, Thought, and Discussion 114

5 Succession Planning 117
Succession Planning as a Hot Topic 119
The Problem. 120
Why We Struggle with Senior Leadership Succession Planning 122
Organizational Value 125
Everyone's Responsibility 126
Making Succession Planning a Cultural Value 127
Farm Team. 128
Dual Role Battalion Chiefs. 130
Sharing What We Learned 132
Preparing Them for the Job: Training. 132
Career Development Planning. 135
Understanding Your Organization's Situation. 136
The Reality of People 138
References. 140
Questions for Further Research, Thought, and Discussion 141

6 Employee Performance Evaluations 147
What Is a Performance Evaluation? 149
Bosses Don't Like Performance Evaluations Either 149
How Did We Get Here: History of Performance Reviews 152
Evaluation Process and the Changing Generations 153
 Changing the Process of Evaluations. 157
Making It Applicable to the Fire Service 158
Dropping the Annual Performance Review 160
Case Study: H.P.F.D. 161
Check-In Meetings for Bosses. 162
Goal Focus. 163
Documentation 163
Some Employers Still Require Annual Reviews 165
Calculating Pay Raises 165
One Size Does Not Fit All 166
Big Picture 179
References. 179
Questions for Further Research, Thought, and Discussion 181

7 Conducting Internal Investigations and Employee Discipline ... 185

- Where to Begin ... 187
- Stopping the Clock ... 187
- Internal Investigations ... 189
- Organizational Culture and Internal Investigations ... 191
- Employee Misconduct ... 192
- Complaint ... 192
- Investigative Process: Preliminary Investigation ... 193
- Investigative Process: Investigation ... 196
- Employee Rights During the Investigation ... 201
- Discipline ... 205
- Documentation ... 212
- Terminations ... 212
- Case Files ... 212
- The Rest of the Story ... 216
- References ... 218
- Questions for Further Research, Thought, and Discussion ... 219

8 Policy Development ... 223

- Guiding Document ... 225
- First Things First ... 227
- Standard Operating Guidelines ... 227
- Categorizing SOGs ... 228
- What Is Included ... 230
- Employee Handbook ... 232
- Legal Validity of Policies ... 233
- Templates ... 234
- What to Include ... 235
- Acknowledgment Forms ... 238
- A Word of Caution ... 239
- Contract Policy Development Firms ... 239
- Job Performance Requirements ... 240
- Time and Motion Studies ... 240
- Conducting the Study ... 243
- Putting It All Together ... 245
- Final Thoughts ... 246
- References ... 246
- Questions for Further Research, Thought, and Discussion ... 249

9 Relationships Matter ... 253

- Relationships and Problems ... 254
- Relationships and the Tenure of Today's Fire Chief/CEO ... 255
- The Role of Relationships and Political Capital ... 257
- Case Study: New Broom Management Theory ... 259
- Relationship with the Boss ... 261
- Relationship with the Elected Board ... 264
- Relationship with the Troops ... 266
- First Call Rule ... 270

| | References. . . . . 272 |
| | Questions for Further Research, Thought, and Discussion . . . . 273 |

10 Wise Counsel and the Weight of the Fifth Bugle 277

Leadership Challenges 279
Kitchen Table Advice 281
Lesson Learned. 282
Command Is Lonely. 282
Wise Counsel. 283
Building Your Network of Wise Counselors 284
Withstanding the Weight of the Fifth Bugle 285
References. 286
Questions for Further Research, Thought, and Discussion 287

11 Epilogue 289

Index. 293
About the Author 313

FOREWORD

Finally! A step-by-step guide for the fire chief's role. New chief, old chief, aspiring chief, volunteer chief, career chief, here it is! In this book, *The Dynamic Fire Chief: Management Theories for Strategic Organizational Leadership*, Chief Craig A. Haigh has put together what the fire service has needed for decades: the tools needed to be successful as a fire chief not just on the fireground or at any other emergency incident but back in the firehouse where, to be honest, we spend most of our time.

There are many who pin on those five trumpets who perform well on the fireground, but when it comes to taking care of the other side of our business that holds it all together, they fall short. Please understand that in most cases it's not their fault. It is a failure within our own system whereby we spend an enormous amount of time preparing the new firefighter for their new profession and justifiably so. We send them to the Fire Academy, EMT school, and possibly paramedic school, and we put them through Hazard Materials and Technical Rescue training and a laundry list of other very-much-needed educational experiences. Eventually we prepare them to be driver operators, teaching and showing them just how invaluable that position is to the success of the team. Next up is lieutenant, then captain, battalion chief, and into a staff chief position of assistant or deputy. Unfortunately, the training and educational experiences offered to those in a chief officer position are few and far between. Some have great mentors; some, sadly, do not. Then one of those chief officers makes it to the big chair—the fire chief's position, chief of department—and they find that the educational opportunities are even more scarce.

Today a fire chief has to be dialed in with ALL aspects of their leadership role—not just the emergencies out on the street but those back behind the bay doors. And consider this: they actually reach out beyond the firehouse doors and into city hall and your county, state, and community as a whole. Some can actually land squarely at the federal level and take on a national spotlight that can be related to grants, funding, and personnel or even a fight concerning an unfunded mandate that can place stresses on an already slim budget. The educated and well-prepared fire chief can perform well when facing these challenges. Those who are not feel the sting of failure.

Today's fire chief needs to be well versed in those political relationships that can at times make or break a plan you may have laid out for apparatus replacement, new firehouses, increases in staffing, and so many more areas that they can't be listed here. In order to be that good "salesman," fire chiefs need to be able to back up what they are pushing for and do so with good solid data and not drama. The stories will not work anymore. Research, data, facts, and a good team are what make your justifications soar and often lead to success. The higher up that career ladder you climb you realize more and more that there is definitely a business side to what we do in the fire service. Those who do not will stumble, fall, and eventually fail.

Chief Craig A. Haigh has provided us all with a gift. A gift that leads to success for you, your team, and most importantly, for those we have sworn to serve. He uses his personal experiences and relates them to each topic. From the complex and complicated, he puts it into a format that we can all understand, Chief Haigh gets it! Those who only use a portion of this book will immediately find success, and it will create a desire within them to be not just good at what they do but great. They'll find themselves rereading this book over and over again, and each time they will gain another advantage as the leader. Each time they will find themselves building on those great accomplishments, and each time they'll become more and more confident within their role, resulting in an unbelievable adrenaline rush, one of knowing you have accomplished something pretty special. That is what Chief Craig A. Haigh can offer to you. It's up to you! Reach out and grab this opportunity, and your legacy will live on forever.

—Rick Lasky, Fire Chief (Ret.)

ACKNOWLEDGMENTS

Special acknowledgment: Robert DeFalkenberg
From speaking multiple languages to your nearly 30 years of experience training managers throughout the world for McDonald's Hamburger University, your expertise in organizational management and business practices has strengthened this book and helped to ensure that my writings and recommendations are more than fire service–focused, but sound business practices overall. Your countless hours of reading, review, and discussion have been invaluable. To my running and exercise buddy and Brother in Christ, I love you, Brother!

To my fire service experts:
From Chief Heim first allowing me to expand my experience with strategic planning, to David Woodward's push to develop a training program for his conference, to discussions with Amy Emerson about the type of book she would find helpful as a study reference for chief officer promotional exams, a book of this nature simply would not be possible without sound review and insight from experts doing the work each day. Each of you have contributed mightily. My deepest thanks!

- Chief Joseph M. Heim, Rock Island Arsenal Fire Department & East Dubuque Fire Department
- Chief Jeffrey Bryant Sr., Amboy Fire Protection District
- Chief John Winters, Silvis Fire Department
- Chief David Johnson, Hampton Fire Rescue
- Training Officer David Woodward, Lake Ozark Fire Protection District
- Elizabeth Woodward, ADN Program Coordinator, State Fair Community College: Eldon Campus
- Amy B. Emerson, Director, Testing for Public Safety, LLC

To my prayer team:
Working more than full-time hours as a fire chief, having my daughter, Melody, get married in the middle of my writing, and then leading the village as their emergency management director through the pandemic of COVID-19 all while trying to write a book in my spare off-duty time was more than a little crazy! Through it all I felt your prayers, gained stamina to continue pushing ahead, and received inspiration, a direction to follow, topics to cover, and stories to tell based on the leadings of our Lord and Savior. This is all due to the prayers of the faithful and your willingness to stand beside me in the gap to do this work.

Ezekiel 22:30

Elizabeth A. Haigh

Melody L. Berendt

Dan Berendt

Leland B. Haigh

Andrea Wilkinson

Robert DeFalkenberg

David Woodward

Elizabeth Woodward

INTRODUCTION

Being fire chief is hard! Whether you are a volunteer, paid-on-call, paid a yearly stipend, or a full-time career chief, the indisputable fact remains that being fire chief is just plain HARD! No matter how many training classes you have attended, how many discussions you have engaged with mentors and fellow chief officers, or how many degrees and certifications you have attained, nothing can fully prepare you for the position. Once you put on the gold badge and bugles, you quickly realize that a whole slew of new topics and concerns now rest squarely on your shoulders.

I remember reading an article back in the early 1990s where the author proclaimed that not all fire officer bugles—a symbol of leadership in the fire service—weigh the same. The author argued that the first, the promotion to lieutenant, and the last, the promotion to chief of department, weigh more than those in between. This does not mean that the physical weight of the bugle is different but rather the metaphorical weight.

The first bugle, in simple terms and at its most basic level, signifies that you are officially responsible. You are responsible for what happens on your watch, for what happens in your firehouse, for what your people do at an emergency incident, and for the way they take care of those you serve. You are responsible to make things operate smoothly and to solve problems and, most importantly, to bring home all the players you "take to the game." As a first level officer (lieutenant), your people and their performance are a direct reflection on you, and you own that responsibility the moment you accept that first bugle.

As you move up the chain of command and gain more bugles, you continue to be responsible, but it is different from that first promotion. The learning curve is not as steep, and the transition not as great. The ranks between lieutenant and fire chief are vital to the overall operation of a fire department. Each rank has different job responsibilities, and each has their own unique set of challenges. But I would suggest that the huge mental transition of initially moving from firefighter to boss does decrease with each subsequent promotion. Sure, you take on more authority as you climb the organizational ladder, you now have new areas of responsibility, and you now supervise supervisors, but the overall initial shock of being an officer lessens as you move up the chain of command.

The fifth bugle however, in my opinion, is in a league all its own. In fact, I would argue that the fifth bugle outweighs all the rest significantly. It still carries all the responsibilities of being a fire officer, as well as some new ones that are unique and can be extraordinarily challenging. It is also a different job, requiring different skills and a different mindset.

The role of today's fire chief, the five-bugle guy or gal, is not that of a "senior firefighter" as so many in the public (and even the fire service) believe. Today's fire chief is really a chief executive officer (CEO) carrying all the same responsibilities one thinks of when hearing that title. They serve in the top executive position and are responsible for ensuring the success of their organization.

A quick Google search will tell you that the CEO oversees the organization's various functions, including compliance, finance, human resources, legal, marketing, operations, and services. They do this with a focus on the needs of constituencies, stakeholders, and employees. They are also responsible for vision casting, producing a quality product/service, and succession planning. Make no mistake; these are all the same functions that a modern-day fire chief oversees as well. We just do it while wearing a uniform, gold badge, and five bugles.

So why do we not generally think of fire chiefs as the organization's CEOs and why do we find so many chiefs who are unprepared to function at that level once promoted? I think the answer is multifaceted.

Whether paid or volunteer, most firefighters entered the fire service with a desire to respond to emergency incidents. As new firefighters we receive countless hours of training to perform skills at the task level. As our careers unfold, we attend classes that focus on tactics, strategies, command and control, and overall incident management. Some of us may take college courses and earn a degree in fire science or a similar field of study. But in the end, most of our training has a primary focus on emergency service delivery. Few courses present on the principles of management needed to effectively run a department. I suggest that since we generally do not view the fire chief position as that of a CEO, we also fail to see the importance of developing CEO-specific skills.

If we look at the education and training focus of today's private sector CEO, we typically see degrees in business administration or maybe finance, law, or human relations. Rising executives focus on how to operate a successful business. Contrast this with the up-and-coming fire chief whose focus is on emergency response, prevention, or training. In general, as a fire service industry, we place little emphasis on learning how to generate and manage revenue, understand employment laws, execute strategic planning, collect and interpret data, or complete a myriad of other CEO-type tasks and skills. Yet this is the job of most fire chiefs. Additionally, the fire chief must engage politically with elected officials and understand how the fire service impacts the demands of their constituents.

The problem goes deeper in that most of us current fire chiefs, including most of our training institutions, do little to teach CEO skills to those who aspire to fill the top spot. We do not talk about the needed skills, do not teach classes that focus on them, and in many ways downplay their importance. Yet it is lack of skill as a CEO that will get fire chiefs into trouble once promoted.

I have seen it repeatedly, a national speaker on fireground tactics and strategy is invited to teach, someone who will show exciting videos of fires burning and firefighters stretching hose and performing their craft, and the class will be filled to overflowing. Because most of us joined the fire service to respond to emergency incidents, we enjoy the classes on response and incident mitigation. It meets our needs as Type A, adrenaline junky personalities.

Now try hosting a class on organizational financial management and strategic planning and see how many students you get. Yet financial management is the lifeblood behind attaining the resources needed to allow effective fireground operations. We as fire service leaders simply fail to make this connection and thereby fail to prepare our future CEOs.

In the end however, I think most fire chiefs, like those of the past, acquired their gold badge and five bugles by being recognized as a good firefighter. These good firefighters, who have superior abilities on the street, often find themselves ill equipped to serve as chief executive officers, not because they want to fail, but because they never prepared to succeed in the job they now have.

The Dynamic Fire Chief: Management Theories for Strategic Organizational Leadership seeks to bridge the gap between service delivery (management of street level operations) and the executive level management needed by a five-bugle-wearing CEO. Regardless of whether you lead a paid or volunteer department, the management skills needed to lead a successful department are the same.

This book should also be viewed as a learning textbook designed to help those rising stars who are seeking formalized leadership to prepare for their future position. Topics covering money management, strategic planning, employee recruitment, hiring, performance reviews, succession planning, and related subjects are covered in an easy-to-understand-and-implement format.

The overarching goal of this book is to help balance the weight of the fifth bugle as well as assist aspiring chief officers in the development of fire service–specific CEO skills that will allow them to manage their organizations with increased efficiency and effectiveness while maintaining a focus on the end goal of providing excellent emergency services.

CHAPTER 1

THE IMPORTANCE OF VISIONING

In 1995 I was both a career firefighter with the city of Rock Island, IL, and fire chief of my hometown volunteer department in Hampton, IL. To top it all off, I had secured an instructor position with the famed Illinois Fire Service Institute (IFSI). At age 27, how could my fire service career be any better?

I was living the life I had always wanted, but strangely I was facing a growing sense of restlessness and discontent. It has taken me years to figure out, but I realize now that this sense of restlessness occurring for no real apparent reason was God preparing me for a major life change. Little did I understand how major of a change He had planned.

As a fluke, I applied for a job I saw announced on the back of a fire department mailer. Following a short assessment center and an interview, I was hired as the first paid fire chief for the King (NC) Volunteer Fire Department. From application to the start of our move, the whole process took less than three months. The governing board was pretty clear on what they wanted me to do—transition an all-volunteer department into a combination agency able to meet the needs of a rapidly growing community just north of Winston-Salem.

My wife, Beth, and I were eager to take on this new challenge. The job would require us to move 800 miles away from family and friends. We knew no one in King and had never even visited North Carolina. The only thing we knew for sure was that we believed that God was calling us to King, and we were happy to follow His lead. Twenty-five years later, we both agree that He was indeed calling us to this place.

WELCOME TO MY NEW REALITY

The job was far more difficult than I had anticipated. To date, it's the most challenging yet exciting experience of my professional life. We developed deep, long-lasting friendships, and we both believe that the change was the absolute best thing we could have done for our marriage. We initially had no one to rely on except each other, which drew us together in a way that I do not think would have been possible without the move. I spent seven years with that wonderful organization and in fact would probably still be there today except that I felt called to return to Illinois to be near our parents.

Although King was well-equipped with apparatus and a new fire station, it had no policies and procedures, no standard operating guidelines, no medical response protocols, no employee handbook, no system on how to test and hire paid firefighters, and no long-range financial system to sustain this newly planned workforce. They lacked a pension system, employer-managed health care plan and had only a very modest leave time policy. So that the department could offer me health insurance, they had cut a deal with a neighboring fire department to place me on their roster so I could receive benefits through their plan. As I look back, I am not even sure this arrangement was legal.

District leadership had decided that they needed to hire paid staff because the volunteer workforce was not able to keep up with the ever-increasing call volume. They had convinced their funding agencies (Stokes and Forsyth Counties along with the city of King) to implement a one-cent tax increase to pay for the cost of a paid fire chief and a full-time firefighter. The problem was that the penny increase would not support anything beyond the two salaries, and the exhausted volunteers needed much more support than a paid administrator and one firefighter.

One other important note—the fire chief and the paid firefighter were expected to answer calls when off-duty as well as work their normal 45-hour work week. The district board had never heard of the Fair Labor Standards Act (FLSA) or exempt and nonexempt employees, and they did not have the funds to pay for any overtime accrued by the full-time firefighter. They were shocked when they learned that we could not work this firefighter in this fashion and that the federal government had rules that covered such things. They quickly figured out, however, that the fire chief was FLSA exempt and that the rule related to hours worked did not apply in my case. I was told repeatedly to do whatever I needed to make the department operate—that meant working almost nonstop.

Depending on the situation, I filled almost all roles within the department, all while wearing five bugles. Sometimes I drove the apparatus, set the pump, pulled the hoseline, and made the attack. Sometimes I forced entry, searched the building, and opened the roof. Sometimes I was the medic providing patient care, sometimes I was the incident commander, sometimes I was the fire investigator, sometimes the rescue diver, most of the time the training officer, and *always* the administrator. We had great volunteers, but the workload exceeded what they could do. Throwing me a bone, the district board agree to give me "hour-for-hour" comp time as an added incentive to my two weeks of paid vacation. In my first four years of employment, I accumulated a little over 2,000 hours of comp time, which meant I did five years' worth of work in a four-year time span. I was exhausted to say the least.

CALLED—BUT NOT YET EQUIPPED

When I arrived in King, I had experience writing standard operating guidelines (SOGs), job performance requirements (JPRs), and a handful of medical protocols. I had also written and conducted officer promotional assessment centers, but I had never messed with all the human resource stuff involved with an employee handbook. I had very limited knowledge of employee pension and investment plans, knew almost nothing about employee insurance, and had a very basic understanding of how to conduct a testing and hiring process for career firefighters. Although I kind of knew how to develop a budget, I had zero understanding of tax rates, assessed valuation, and revenue generation. I had no college degree, had never attended classes or trainings on such things, and also had no clue about politics in a small southern community. I found myself living out a quote by author Rick Yancey: "God doesn't call the equipped, son. God equips the called. And you have been called" (Yancey, 2013, p. 131). I learned quickly that my role in living out this calling was to be teachable. I needed to figure out what I did not know and then learn it—as quickly as possible.

FIGURING IT OUT

I can't count the number of times that I have assigned someone a project only to have them return to my office after a few weeks of spinning their wheels to report that they don't know what to do because they have never been trained or attended a class. They tell me that I either need to find someone else to do the work or we need to abandon the project altogether.

There is not always a class, certification, or degree program to train you to do something. I am a huge proponent of formalized education. But beyond all else, you need to learn to think. This involves learning to evaluate, reason, and cast a vision for what needs to be done. Innovation comes from vision, and vision comes from being able to see what is possible. The Wright brothers did not develop manned air flight because they went to a class that taught them how to build an airplane. NASA did not put astronauts on the moon because they sent their engineers and scientists to classes on how to build a manned rocket ship. Doctors did not learn how to do a human heart transplant by finding a "transplant class." They had vision for what needed be done . . . and they figured it out (Haigh, 2019).

Many times over the years I have had to just "figure it out." I have been asked repeatedly by my superiors to work on projects, build programs, or do things that I have not been trained to do or had experience with. In every one of those cases, it was not just figure it out—but produce work with excellence. If you want to be successful as a fire chief/CEO (or really anything in life), you need to be willing to put in the hard work to figure it out. This was the lesson I was learning in King and one that has served me well throughout the rest of my career.

WHERE TO START?

Looking at the massive amount of work in front of me I felt like Gutzon Borglum starting his carving work on Mount Rushmore (U.S. Park Service, 2020). I had a rough idea in my mind of what the department would look like once all the parts and pieces came together, but creating an organization that would ultimately be a quality employer, able to attract the best and brightest, while continually meeting the ever-changing emergency response needs of the community was overwhelming. I also realized that the project was not going to be a quick process and that it would take years to fully develop and construct. Just as Borglum had created a miniature model of what the mountain would look like when complete, I quickly learned that the only real path forward was for me to sketch out the scope of the project and then to break it up into manageable parts. To do that, I needed to spend some time visioning.

VISIONING

Visioning is an essential part of managing an organization. Visioning helps you see what is possible so you can develop a directional plan. It involves a variety of linked steps such as strategic planning, financial management and succession planning, but the most important step in the process is taking the time to think. Thinking critically about your organization involves assessing how well it is meeting its overall mission. It also includes looking at the action steps required to prepare, strengthen, and build it for the future. In my visioning process I like to use these two distinct yet intertwined steps:

1. I allow my mind's eye to imagine what the future looks like.
2. I develop a construction blueprint that when built creates the structural foundation for what I imagined. (When creating blueprints, do what works for you. I have developed some of my best blueprints on paper napkins during lunch.)

Visioning is a proactive process. It is the absolute antithesis of reacting. When we as leaders do not take the time to think and to establish a vision, we by default become reactive. Being reactive in the absence of visioning yields distraction. Distracted leaders then lose focus on what is most important—the overall mission of the organization and taking care of the work force needed to manage the mission. In my experience, the key ingredients in the "figuring it out" recipe is working to limit distractions, avoiding reactive decision-making, and then thinking.

LEARNING FROM OTHERS

When beginning to think about your organization, it is important to consider that somebody else (in another organization) has probably been in a similar situation. Whether you are assessing your current circumstances, evaluating the steps required to fix a wicked problem, looking to build a program, or working to construct a foundational structure that will be needed in the future, there is a strong likelihood that some other leader has probably already done, at least in part, what you are looking to do.

I am a huge fan of case studies. I learn best through stories that tell tales of success and failures and how others have led through challenging situations and conditions. "A case study is a research approach that is used to generate an in-depth, multi-faceted understanding of a complex issue in its real-life context. The value of case studies is well recognized in the fields of business, law, and [public] policy" (Crowe et al., 2011). In my opinion, becoming a student of case studies is an important part of organizational (as well as life) visioning.

Case studies are a form of storytelling. I personally am attracted to people who can convey a message through storytelling that compels you to think about how the story applies to your organization or maybe even your life. I especially like those who are transparent and comfortable in their own skin—warts and all. I value this transparency because it eliminates pretenses and tells the good, the bad, and the ugly to help someone else grow. It is not necessary that all aspects of their story align with yours—it just needs to get you to think, because thinking is the key part of visioning.

Storytelling (i.e., case studies) is such a big deal that the National Fire Academy's class on executive leadership identifies "storytelling" as an essential part of leadership (Haigh, 2017). Storytelling capitalizes on the fact that most of us learn and can recall information presented in a story format better compared with simply memorizing facts and figures (Denning, 2004). This is the exact reason we use nursery rhymes to teach children.

Think about the story of the three little pigs. It contains all parts of a great story—characters, setting, plot, conflict, and resolution. It may have been years since you heard or thought about the story, but my suspicion is that you likely remember the key parts—the personalities of each of the pigs (lazy, less lazy, and a hardworking), the materials the pigs used to construct their houses (straw, sticks, and bricks), the wolf (antagonist), and how the pigs were saved (by the protagonist's hard work). The following takeaway messages are conveyed through the story:

1. Hard work and dedication pay off.
2. When trouble comes, and it will, preparation pays big dividends.
3. For our public safety personnel—be prepared to lend a hand and help someone else when they are in danger (i.e., let the other two pigs seek shelter in the protection you provide).

No child would recite these points, although they are the message of the story. I bet however that we all remember the story, and it is likely that at some point and time in our own lives, we have each applied these principles.

FINDING CASE STUDIES

To be effective in learning from others, you need to have relationships. You need to have interaction, sharing, and as the late Battalion Chief Tom Hatzold, my former chief of inspectional services, used to say, "time to visit." It requires that leaders put themselves in situations where they can do these things. Sometimes it involves actually traveling and looking at other departments, watching others do things, attending trainings, going to meetings, taking time to share a meal with your peers, being available for a phone conversation, listening, reading about what others are doing, and always being willing to share your work product in order to help someone else improve or solve a problem. Gaining information to be used in visioning is much more than sending out a survey through one of the many professional organizations. It is connectedness with others so that you get to hear their stories and learn from their challenges. You then catalog this information so you know who to call when you need help. As they say here in Chicagoland, "I got a guy!"

THE UNTOUCHABLES

For many Chicagoland public safety professionals, the actions of Special Agent Eliot Ness are legendary. U.S. Attorney George E.Q. Johnson chose 27-year-old Ness to lead a small squad that became famous for taking down the reputed bootlegger and crime boss Al Capone. Due to his notoriety, Ness became the model for Chester Gould's cartoon detective Dick Tracy (Bureau of Alcohol, Tobacco, Firearms and Explosives, 2020).

Important to recognize from this story—Ness could not have done the work alone. He was only able to accomplish the mission because he was able to convey a shared vision to a small group of dedicated team members. He was able to paint this vision in such a way that the team felt it worthwhile to put their full efforts, including their lives, behind accomplishment of the vision.

MY OWN GROUP OF UNTOUCHABLES

In King, I needed help, and I knew that I could not accomplish what I had been hired to do without a few others sharing the vision. When I arrived, about half of the department did not see the need and did not want to hire paid staff—therefore my being there was a very visible point of contention. The other half wanted paid folks, but many had underlying self-focused intentions about either becoming one of the future paid personnel or getting one of their friends hired in these new roles. This was coupled with the fact that I was an outsider—a Yankee—a fact that I really did not understand but one that certainly made the work to be done even more challenging.

Prior to me being hired as the first paid fire chief, the department's board of directors annually selected and appointed, or reappointed as the case might be, the department's fire chief. The newly appointed chief then hand-picked their officers. The process frankly did not work very well and usually resulted in hard feelings and overall organizational dysfunction. As an example, a member could spend a year serving as an assistant fire chief (or any other officer rank) and then following the annual meeting find themselves the next day back at the rank of firefighter.

Likewise, another department member who had previously served at a firefighter's rank could immediately switch roles and now be an assistant chief (or any other variation of officer rank). This system resulted in zero consistency. All rules, regulations, standards, and past practices were up in the air and basically null and void as soon as the vote of the board was announced.

As the new chief, it was my responsibility to select my officers (they had all been demoted by my appointment). The problem was, I did not know any of these people or their abilities. As a compromise (really a necessity), I decided to keep everyone in the position they held prior to my appointment while I conducted a promotional assessment center to find the candidates most qualified for each rank. As I explained to both the membership and the board of directors, under the new structure we were building, this assessment center promotional process would become policy within the employee handbook for how officers would be selected going forward. The rotation of officers would be eliminated, and officers would receive permanent appointments following a probationary period. They would then hold their rank unless promoted again or demoted due to disciplinary action.

Two key leaders emerged from this promotional assessment center. Michael Merritt, one of the most talented fire officers I have ever met, was promoted to deputy fire chief (second in command), and V. Keith Handy was promoted to lieutenant. Both officers worked tirelessly to help me set the direction and build the foundation for where we needed to go. I quickly discovered that both individuals are truly brilliant leaders, and their dedication was nothing short of amazing. I hired Lieutenant Handy quickly after the assessment center to fill the full-time position the board had planned. This made him the department's second full-time employee. With Lieutenant Handy working Monday through Friday and Chief Merritt spending most evenings at the fire station (after he got off from his "real job"—Chief Merritt was a volunteer) we were able to jointly begin developing the policies needed to transition the department into a combination organization. Chief Merritt now has more than 45 years of service to the department, and Keith Handy now serves as the director of planning and inspections for the city of King.

Another one of my Untouchables was my secretary, Barbara Thorpe, and her husband, Ron. Barb's job title could have easily been changed to "director of intelligence" rather than "administrative assistant." Barb worked hard to maintain the pulse of the membership and was fantastic at carrying my message to the membership and select members of the community. Remember—the message ALWAYS matters. Her husband, Ron, using his expertise in information technology and the private business sector, helped us as an organization move from paper and pen into the digital world. He was also a great sounding board and is an individual with great wisdom.

Rounding out this team was Board President Brent Davis and his predecessor Robert Woods, as well as our accountant, Barry Amburn, who not only watched the money but was my confidant and prayer partner, and my great friend Scott Haithcock, who served as my constant encourager and who often felt like my personal bodyguard and protector.

Lastly, and the toughest fighter of them all—my wife, Beth. The role of your partner cannot be overemphasized. Through their support they will either push you to succeed or, by their lack of support, hold you back or, worse yet, cause you to fail. In my case, Beth was and still is my support system, my encourager, and the one who constantly tells me to take the risk, reminding me that it is worth it. She is also a natural at "representing me," while I "represent the organization." The importance of this last trait cannot be overemphasized.

The following quote is often attributed to the Greek philosopher Heraclitus:

> Out of every 100 men [in battle], ten shouldn't even be there, eighty are just targets, nine are the real fighters, and we are lucky to have them, for they make the battle. Ah, but the one, one is a warrior, and he will bring the others back.

Reflecting on this statement, especially as it applies to emergency services and local government, I believe that it is as true today as it was then. I did not know when I took the job in King that I was being called to be a warrior. In fact, if I had known, I would not have moved my family to King out of my own insecurities and fears. But I did not know—I was just following God's lead. I also quickly recognized that no warrior succeeds without a group of real fighters. The fighters mentioned here were my group of Untouchables. They had caught my vision and were willing to fight to make it happen.

WHY WE STRUGGLE WITH VISION

Why do we struggle with visioning? The district board in King had a vision of how to better meet the needs of their community—they were not exactly sure how to do it, but they had a vision. Our nation's leadership had a vision of cleaning up corruption in Chicago—they did not appreciate the cost, but they had a vision (i.e., they appointed Elliot Ness). Dr. Christiaan Barnard, who transplanted the first human heart, did not know that he would be accused of murder for removing the beating heart from a brain-dead patient, but he had a vision.

I think the answer to this question is that we instinctively know that risk exists anytime we try and implement a vision. We may not understand the depth of the risk, but we know that it is there. We as individuals are afraid of risk—therefore, knowing that visioning will cause risk, we simply do not work to cast a vision. However, without vision, the danger is actually greater.

EVALUATING YOUR WHY

Because of my experiences and my willingness to share what I have learned, with unfettered transparency, through writing, speaking, and teaching, fire officers from across the nation contact me to talk about their interest in becoming a fire chief/CEO. During these conversations, I almost always ask about their vision and what they want to accomplish if they receive the promotion. I am interested in their why. *Why* do they want to be fire chief/CEO? *Why* do they want to take on the added weight of the fifth bugle?

The why varies from individual to individual. Some are interested in padding their pension with a higher rate of pay. Some see the position as the pinnacle achievement of their career. Some see it as "their turn," but occasionally I talk to someone who wants to do something great and has a vision to make things better.

I believe that being fire chief is a fantastic job and that having the ability to set the course for how an organization meets the needs of those it serves is truly an incredible experience. But I also want those interested in the position to understand that accomplishing a vision will likely be much harder than they anticipate. Are they willing to take the risk? In today's litigious environment are they willing to put their house and their pension on the line to accomplish their vision? How about the cost to their family, loved ones, and their overall reputation as an officer and as a person? Usually they ask, based on my experiences, would I do it again? The answer is unequivocally—Yes!

TIME DEMANDS

If we can get beyond the fear of visioning, we will assuredly be met with the challenge of time. Back in 2013 I wrote a class called Juggling Hats for Winter Fire College at the University of Missouri—Fire and Rescue Training Institute. This class asked the following question: "As a fire service leader, how many different hats will you be required to wear in a single day?" The question I was really asking was how much can you get accomplished within the time you have available? (Haigh, 2013) (table 1–1).

Whether you are paid or volunteer, the passion to serve can quickly be overshadowed by the demands and the simple fatigue associated with this

Table 1–1. What are your jobs and what can you accomplish within your time?

Hats to wear...	Time demands...
• Emergency responder	• An increase in customer expectations
• Accountant	• Increases in call volume
• Human resource director	• Unfunded mandates
• Legal expert	• Flat or declining revenue
• Politician	• Greater training and certification demands
• Cheerleader	
• Marriage counselor	• Staff availability
• Boss	• Recruitment and retention

workload. In the years since I wrote that original class, the situation has not improved. Today's fast-paced and demanding environment causes the workload on the fire chief/CEO to be incredibly challenging:

- 24-hour connectedness (smartphone, social media, email, text messages)
- Office hours (constant demand to be producing)
- Lack of time to read and study
- Meetings, appearances, public events, political events, and so forth
- Lack of staffing (especially in the administrative roles)
- NO time away from the job—it has become 24/7/365

If I were asked to put a letter grade on how many of us function today, I would say that we are a B– industry with constituent expectations demanding an A+ performance. To produce A+ work we need time to vision, but when there simply are not enough hours in a day and the fatigue becomes overwhelming, we do our best to stay one step ahead of the hungry pack of lions known as elected officials, the media, and our own constituents. When you are simply running to survive there is no time to think about visioning.

THE JEALOUS MISTRESS THAT IS KILLING US...

I have started referring to the job as the jealous mistress. It so consumes our lives that we have room for nothing else. Our mistress is jealous of *anything* that impacts the time we should be giving to her. I cannot count the number of fire service and government leaders who we are losing today, not because they are ready to walk away and certainly not because they have nothing left to contribute but because the job is making them mentally and physically sick.

The big question of our day: Do those we work for and those we serve care what their expectations are doing to those who lead their emergency service organizations? Bigger yet, how does this situation impact our ability to vision for the future?

DON'T GO IT ALONE

Although I talk about visioning as a primarily individual process, I also think it is important to include others in visioning. Talking, sharing ideas, hearing what others think, and learning what their personal vision is for your organization is a huge part of getting it right. I believe getting away from the office and limiting distractions is a paramount step in this process.

Since I continued to teach at the Illinois Fire Service Institute while I worked in North Carolina, almost every time I made the trip back to Illinois, I brought along some of my King personnel. Twelve hours together in a vehicle along with the sharing of meals and hotel rooms gave us an incredible amount of time to talk, think, and plan. Those who came along attended training classes, talked to others in the fire service, and were exposed to some of the best instructors the U.S. fire service has to offer. I strongly believe that this helped to build the King Fire Department into the great organization that it is today.

I know of another department that takes their command staff as well as their spouses away for a working retreat. During the day the command staff meet and work on team visioning, and then everyone comes back together for dinner and after-hours activities. What a great opportunity to build teamwork and set the vision for your organization.

GONE ~~FISHING~~ VISIONING . . .

The reality is that visioning is expected and required but typically not supported by our public sector employers. What would happen if you told your boss that you were leaving the office and taking the afternoon to think and work on visioning? How about taking two days away to do the same thing or spending dollars to take your command staff and their spouses away for a working retreat? Does your boss trust you enough to believe that you are actually working? What if someone asks where you are? How do they justify to the taxpayers that you are out of the office thinking? How does this play with the perception of wasteful and inefficient government workers and spending?

As I think about how to move forward related to this issue and the lack of visioning time, several questions come to mind:

- Is this a new and emerging issue in today's work environment?
- What will the future hold as it relates to this issue?
- How will it impact potential future leaders related to their desire and willingness to pursue senior management positions?
- How will we develop solutions if time to think is not supported?

Here is an assignment:

> You are tasked by the City Manager to develop a recommendation to address the concerns related to "thinking/organizational visioning time." This recommendation is to be officially adopted as policy by your organization and must be approved by the governing board.

What would your recommendations be?

If you want to be on top of what I think is an emerging issue in today's marketplace of government service, start thinking about and visioning a solution for these questions.

THE MESSAGE MATTERS

Once you find the time to vision and you develop a plan for where you need to head, a word of caution as it relates to messaging your vision: Often the message we believe that we send to our bosses, employees, stakeholders, and constituents is not the message that they receive or hear. Interpretation of the message, based on the personal biases and beliefs of the receiver, always has the potential for confusion or lack of clear understanding. This confusion can plague projects and have a negative impact on the organization.

MESSAGING MISCONCEPTIONS

As a fire chief/CEO we need to continually be assessing how a message is perceived compared against the one we intend to send. As an example, a department back in the mid-1980s, let's call them the Camp Leland Fire Department, developed a vision to begin providing the 9-1-1 ambulance service to their community. Camp Leland is an old city with a well-established and high-quality fire department. Their vision was sound, and the plan would do much to increase the level of EMS service provided.

Working together, fire department leadership along with the union representing Camp Leland shared their vision with city leadership. The plan was that the department would temporarily reassign personnel from their ladder companies, thereby making these units unstaffed, so they could then staff the new ambulances. The plan clearly said that this move would be a temporary staffing change in that as soon as the ambulances started producing revenue the city would hire additional firefighters so the department could return to having staffed ladder companies.

The message heard by the elected officials however was slightly different than the one the department intended to send. The message received was that ladder companies are not necessary and ambulances generate revenue. Fast forward almost 40 years, and the department still operates using this "temporary" staffing model. In the end, we as leaders need to understand and recognize that our message is not always heard or interpreted correctly, and our failure in messaging our vision can have long-term consequences.

Over the years I have found it incredibly helpful when developing the message of my vision to pull together a group of confidants and ask them to shoot holes in my message. I ask them specifically to look around corners to see what I am missing and to challenge me to make sure that I am thinking correctly. I have avoided more disasters through this process than I can count!

AVOIDING THE UNDERWATER ROCKS

No organization can remain stagnant. Those who are not progressing and moving forward or who are trying to simply not lose ground are failing those they serve. It may appear from the stagnant leader's vantage point that all is okay, while a broader view will show the thunderclouds just off in the distance. As a leader it is important to understand the constantly changing landscape of emergency services and be working to address challenges before they become a crisis. Every organization, regardless of size and geographical location, is undergoing changes.

Our world is constantly changing, as are the demands of our constituents and our workforce. Visioning is the tool that prevents the need for reactive management. Being progressive involves becoming a student of our business. This includes constant learning, thinking, and watching for variations or blips in your data that might be predictive of changing conditions. When you hear firefighters talk about a progressive fire chief/CEO, what they are really saying is that the leader is doing well related to visioning.

Visioning is hard, and I would argue that it is getting more difficult based on the changing dynamics of how we work. As a leader it is not an option for us to pretend that change is not occurring and to try and ride out our time until retirement by simply hoping nothing blows up. As the leader, you are the captain of the ship, and you need to steer it through the water avoiding the unseen rocks and obstacles to prevent damage. A hole in your organizational boat is a bad thing!

Likewise, it is not okay as a leader to kill your employees by demanding that they bail water out of your sinking boat (i.e., your organization) while it is taking on water faster than they can throw it out. Some will tell you that the obvious priority is to fix the hole. Fixing the hole is being reactive. Although this work may absolutely be required, sound visioning will prevent you from hitting the rock that caused the hole in the first place. Preventing a hole in your organizational boat is always better than trying to fix a hole!

I can only speculate on where emergency services are headed in the future. However, I can say with absolute certainty that the only way we as leaders are going to be able to address what is coming in a nonreactive fashion is by using all the tools at our disposal so we can obtain the information needed to allow development of a vision for the future. Visioning is thinking strategically, and strategic leadership is an absolute must for an organization to be successful.

REFERENCES

Bureau of Alcohol, Tobacco, Firearms and Explosives. (2020, July 14). History—Eliot Ness. https://www.atf.gov/our-history/eliot-ness.

Crowe, S., Cresswell, K., Robertson, A., Huby, G., Avery, A., & Sheikh, A. (2011, June 27). *BMC Medical Research Mehodology,* 11, 100. https://www.ncbi.nlm.nih.gov/pmc/articles/PMC3141799/.

Denning, S. (2004, May). Telling Tales. *Harvard Business Review, 82*(5), 122–129.

Haigh, C. A. (2013). *Juggling Hats: The Multiple Responsibilities of Managing a Vol./Combination Department.* Columbia, MO: University of Missouri—Fire and Rescue Training Institute: Winter Fire College.

Haigh, C. A. (2017). *Training Class: Leadership Principles for the Fire Officer.* Champaign, IL: University of Illinois—Fire Service Institute.

Haigh, C. A. (2019). Opportunities / Figure it Out / It is Personal. *Revolutionary Fire Tactics at the Lake.* Lake Ozark, MO.

U.S. Park Service. (2020). *Sculptor Gutzon Borglum.* https://www.nps.gov/moru/learn/historyculture/gutzon-borglum.htm.

Yancey, R. (2013). *The 5th Wave.* New York: The Penguin Group.

QUESTIONS FOR FURTHER RESEARCH, THOUGHT, AND DISCUSSION

1. How does the Fair Labor Standards Act (FLSA) define the difference between an exempt and a nonexempt employee?
   _____
   _____
   _____

   a. What is the FLSA exemption test, and how does it apply to the employees of your organization? Which employees fall into each category and why?
      _____
      _____
      _____

   b. What are the compensatory time accrual and use requirements for FLSA *exempt* employees?

      i. For exempt employees, does compensatory time accrue at a straight rate or at a time-and-a-half overtime rate?
         _____
         _____
         _____

      ii. Is there a maximum number of hours that can be accrued?
         _____
         _____
         _____

      iii. Does compensatory time need to be used within a specific time frame?
         _____
         _____
         _____

2. If you were tasked with creating a new emergency service organization from scratch, what would it look like related to employee pay and benefits?

   a. What employee benefits would you consider essential?

   _____
   _____
   _____

   b. What metrics would you use to develop your employee pay plan?

   _____
   _____
   _____

3. If asked to apply a critical eye to your own organization, how well is it currently doing in meeting its overall mission?

   a. To its constituents?

   _____
   _____
   _____

   b. To its employees?

   _____
   _____
   _____

4. What do you view as the five greatest needs facing your organization and why? What role will you play in addressing each?

   a. _____
   b. _____
   c. _____
   d. _____
   e. _____

5. Who are your personal "top five go-to resource people" when looking for answers to help you lead your organization? Why?

   a. _____
   b. _____
   c. _____
   d. _____
   e. _____

6. Describe your personal WHY.

   _____
   _____
   _____

7. Describe ways within your personal organization that you can be intentional about creating opportunities where your team can meet, with limited distractions, to focus on organizational visioning.

   _____
   _____
   _____

8. You are tasked by the City/Village/County/Township Manager to develop a recommendation to address concerns related to "visioning time." This recommendation is to be officially adopted as policy by your organization and must be approved by the governing board. What would your recommendations be?

   _____
   _____
   _____

9. From your personal experience, provide an example where your message was received differently than intended and the associated impact? What did you learn from this experience?

   _____
   _____
   _____

CHAPTER 2

STRATEGIC PLANNING

Strategic planning is the formalized written process of visioning. It works to set an action plan to accomplish an organization's vision and is a tool to prevent reactive organizational management.

EAST DUBUQUE EXAMPLE

East Dubuque is a volunteer fire department that pulled off the seemingly impossible. They created and operated a fire-based, Illinois-licensed paramedic ambulance service using an all-volunteer workforce. Illinois EMS rules back in the early 1990s related to paramedic staffing requirements were as tough and rigid as a long spine board. But somehow East Dubuque, the tiny fire department located in the far northwest corner of Illinois along the Mississippi River and just south of the Wisconsin state line, figured it out and made it happen. They were heralded in Illinois as an example of what was possible with strong visionary leadership and a motivated group of dedicated volunteers. In the years before I left to head to North Carolina, I got to know a few of their volunteers, and I often picked their brains to see what they were doing and how they were making it all work.

When I returned to Illinois in 2002 after serving in North Carolina, I was saddened to learn that their volunteer paramedic program had been shut down and that they had been forced to contract with a private ambulance provider. They had fallen victim, like so many others, to the national plight of limited volunteer availability.

Shortly after returning to Illinois and taking up my duties in Hanover Park, I was asked by the University of Illinois Fire Service Institute (IFSI) to travel to East Dubuque and assist with a live burn training in an acquired structure. In setting up and conducting the training, I was blessed to meet Chief Joseph M. Heim. Chief Heim is a true professional, serving both as the paid fire chief of the Department of Defense—Rock Island Arsenal Fire Department and as the volunteer chief of East Dubuque Fire Department. He is a visionary leader who loves the fire service—both career and volunteer. He is smart, able to think and reason, and dedicated to excellence, and he sees zero difference in the level of professionalism expected and demanded from fire service personnel, regardless of whether they volunteer or receive a paycheck. He and I became fast friends (fig. 2–1).

Flash forward to 2011. Chief Heim had just listened to me speak on strategic planning at FDIC, and as soon as I finished, he approached me and asked for my assistance in developing a plan for East Dubuque.

Figure 2–1. Chief Joseph M. Heim standing next to one of the props at the East Dubuque (IL) training facility (photo courtesy of East Dubuque Fire Department).

LEARNING THE IMPORTANCE OF PLANNING

When I started my career back in 1983 as a volunteer in Hampton, we did not have a lot of fancy resources or specialized tools. Our budgets were small, and pancake breakfasts and chili supper fundraisers were a necessary mainstay of our operation. Funds were so tight that planning was essential. We did not have money to spend unwisely, so we scrutinized every decision to ensure that we were being the best stewards of the taxpayers' hard-earned dollars.

Chief Larry L. Anderson was a well-respected and regionally known professional with marketplace management skills developed through his career job with 3M. He was a second-generation volunteer firefighter at Hampton and modeled for me the example of how to be a fire chief. Chief Anderson annually led his leadership team through the process of planning and project status reviews before strategic planning became popular in government. As a young firefighter, I saw the benefits of community risk assessment, budget planning, revenue generation, and careful project management. I felt the excitement build with the accomplishment of each small objective and the pride associated with bringing a long-sought-after goal to fruition. He taught me to work hard, keep my eye on the ball, and not be distracted by the latest fad or what cool new tool our well-to-do neighbors just purchased. I realized the importance of strategic planning and its impact on creating organizational excellence. Without this experience, I may never have discovered how important planning is for organizational success (Haigh, 2016).

IT IS NOT THAT DIFFICULT

Do not be afraid of strategic planning. It really is not that difficult. It requires thinking and writing a few things down. Prior to my retirement, I was responsible for the management of the Village of Hanover Park's comprehensive strategic plan (both organizational and operational). Even though I managed the process for many years, I have not had any specialized training on the topic and am pretty much self-taught. I watched Chief Anderson, have read a couple books on the topic, listened to a lecture at the National Fire Academy, and have reviewed as many plans from various organizations as possible—but I have never been formally trained on how to do strategic planning. I guess you can say I just figured it out. Honestly, it has always seemed to me to be a process with a nebulous sounding title that is pretty much plain common sense. Also, the more you do it, the better you become. So, when Chief Heim asked for my help and extended an offer to interact with the legendary East Dubuque Fire Department, I jumped at the chance.

SETTING A HIGH STANDARD

As we started the planning process, Chief Heim made it clear that he was interested in moving his department to the next level and felt strategic planning would provide a clear road map to guide the department to future success. As he described it, "East Dubuque Fire Department is a good department, but it is not a great department." Chief Heim had a vision to grow the department into a great department and recognized the role of strategic planning in the overall process.

In his book, *Good to Great*, Jim Collins (2001, p. 213) explained that there are so few great institutions because there are so many good institutions:

> We don't have great schools, principally because we have good schools. We don't have great government, principally because we have good government. Few people attain great lives, in large part because it is just so easy to settle for a good life. The vast majority of companies never become great, precisely because the vast majority become quite good—and that is their main problem.

Chief Heim wanted his department to be great! So we set out on the journey together. I asked his team to think critically, analyze, and set a vision while Chief Heim did the hard work of avoiding the underwater rocks that could sink the process.

IDENTIFIED NEEDS

Getting started, Chief Heim brought together key leaders of the department representing all ranks and levels. Members of the city council as well as the city manager, the police chief, and the public works director all participated in the initial meetings. Those participating were selected based on their organizational involvement, expertise, and need to support the plan once it was complete. By assembling this diverse group, discussions were able to occur not only on the small issues of needed equipment or training but on the much larger concerns of finance, community growth, and the needs of all city departments, including ways to reduce silos and encourage more cross-department interaction and sharing.

The end goal was to develop a plan that would be used daily to guide overall decision-making. Another key aspect was that the plan was to be viewed as a living document and not be so restrictive in that it could not be adjusted and realigned to meet the changing demands of a dynamic municipal environment.

Five critical issues came into focus after countless hours of discussion and analysis:

1. Limited available revenue was stifling the department. The city was just not able to garner enough dollars to fully sustain the department's needs, which was thereby causing them to fall further and further behind financially with each passing year.
2. The city and the department were exceedingly unhappy with their private contracted ambulance provider who they felt was not adequately meeting the needs of their constituents. As with most private services, the contractor needed to balance nonemergency patient transfers against the availability to respond to 9-1-1 calls. This balancing act is tremendously difficult, and East Dubuque felt that the often delayed response times experienced by their community were simply unacceptable.
3. They had a need for a live burn training facility to maintain the skills as well as the energy and excitement of their volunteer workforce.
4. They had some serious apparatus replacement needs.
5. Their fire stations were simply too small to house modern apparatus.

As this list came together, I remember thinking that their needs exceeded what could reasonably be addressed in a measurable time frame. It would likely take decades of hard work to address all these issues. What I failed to realize was that the same passion and drive that had built the paramedic ambulance service years before was still woven into the organizational fabric of the department today—I simply underestimated them. I was failing to recognize organizational and personal grit.

GRIT

In the 2018 article "Organizational Grit," published in the *Harvard Business Review*, the authors wrote,

> High achievers have extraordinary stamina. Even if they're already at the top of their game, they're always striving to improve. Even if their work requires sacrifice, they remain in love with what they do. Even when easier paths beckon, their commitment is steadfast. We call this remarkable combination of strengths "grit."
>
> For leaders, building a gritty culture begins with selecting and developing gritty individuals. What should organizations look for? The two

critical components of grit are passion and perseverance. Passion comes from intrinsic interest in your craft and from a sense of purpose—the conviction that your work is meaningful and helps others. Perseverance takes the form of resilience in the face of adversity as well as unwavering devotion to continuous improvement. (Lee & Duckworth, 2018)

East Dubuque Fire Department is a gritty organization with an exceptionally resourceful leader! Any leader can have vision, but only the truly exceptional leaders can establish an organizational culture where team members embrace the vision, make it their own, and then put in tireless effort to bring the vision to fruition. But this is what you get when you deal with Chief Heim and the East Dubuque Fire Department. Whether it is a member who can squeeze grant money out of a turnip (they have one of those members) or one who can build and fabricate an end product simply based on an idea scrawled on a napkin (they have one of those members too), these unique member skills are harnessed by Chief Heim to make it all happen.

PLANNING AND EXECUTION

After carefully analyzing their situation, we dug in and wrote some specific goals and associated objectives to address their challenges. We established deadlines and assigned overall responsibility for specific tasks to specific people who would champion the mission and get the work done. The initial strategic plan was beyond aggressive, and as I drove home after presenting the final document to the leadership team, I remember thinking that they had no chance of getting it all done. Boy, was I wrong!

Within months they had created and implemented a cost recovery program. Putting this new system into operation required the development of a fee structure, the hiring of a billing firm specializing in fire and EMS services, and getting the whole process approved by the city council. Suddenly, a previously untapped revenue stream based on insurance claims developed, and money started flowing in from insurance companies paying them for responding to calls and providing services to their policy holders.

Next, they tackled their ambulance problem. They convinced the city leadership to reenter the ambulance business and designed a plan to make it financially self-sustaining based on billing. They started an aggressive member recruitment campaign, purchased a preowned ambulance, renewed their EMS Transport and Paramedic License, and dissolved the contract with their private provider. The new service opened for business July 2014, and the team has not looked back. They even replaced their used ambulance with a new one using funds generated through billing.

Work then started on the training site. We had carefully outlined a series of steps that, if completed, would allow the department to begin construction by the spring of 2016. But true to form, on October 11, 2015, far ahead of what had been planned, I was honored to serve as the keynote speaker for the dedication of the Chief Joseph M. Heim Training Grounds, a name selected by the membership to honor their visionary and gritty leader.

As time has passed, we have developed subsequent strategic plans, each one building upon the success of the last. As of this writing, the department is in the process of completing a station design study with the intent of replacing their headquarters fire station and updating their substation. As their aggressive work continues, I have come to view them as the poster child for what is possible with sound strategic planning.

WE ALL MAKE PLANS

Think of it like this.... We all make plans. Plans can be as simple as contemplating what to have for dinner to planning your weekend activities. It could be planning for a vacation, a wedding, paying off a mortgage, saving for retirement, or a myriad of other topics and situations that cause us to think and focus on how best to accomplish a goal. Benjamin Franklin, the nation's first fire chief, supposedly once said, "If you fail to plan, you are planning to fail" (Bouffard, 2011). Therefore, using the words of Chief Franklin, one could say that planning is important.

MAKING LISTS

Part of planning often involves making lists. We make grocery lists, to-do lists, and lists that help us remember the step-by-step processes involved with various tasks. We develop flow charts and cheat sheets to help us remember complicated processes. Every paramedic is familiar with patient care protocols for cardiac mega codes and the algorithms designed to help us remember medication dosages, administration rates, and the associated timing between defibrillations. Based on our extensive use of lists, one could argue that these lists are the tools used to order the steps of planning.

Strategic planning, at its most basic level, really is nothing more than the process of planning what you want to do and then making a list of the steps needed to get it done. Also, just as my wife routinely asks me what I want added to the grocery list, we should probably ask those who use our services (both our internal and external customers) what they think needs to be added to our organizational planning lists.

EYE FOR THE FUTURE

When the term strategic planning is used in the context of organizational leadership, people often mistakenly try and equate the process to an attempt to predict the future. Rather, it is a well-thought-out and deliberate study that looks at the decisions we make today and how they will likely impact the organization's future (Haigh, 2016; Wallace, 2006).

Strategic planning also asks an organization to evaluate what it is, what it does, and whether it is really meeting the needs of its community (Haigh, 2010). By looking at these questions, the answers help to clarify the organization's mission and thereby provide guidance and direction in choosing the right path for future success (Wallace, 2006). The process basically revolves around these fundamental questions:

- What does our community need us to do?
- What is needed today, and what will likely be needed in the future?
- What would a blueprint for action look like?
- How will we know if we are on track?

EXAMPLE: REPLACING A WATER MAIN

Think about a community that is planning to replace a section of aged water main. Based on the specific location of the main, the surrounding property, and the availability of roadway access, the community may want to consider increasing the size of the main to support future economic development. Water mains are expensive, and it would be far cheaper to upsize the pipe while the workers and the equipment are on-site and available than to replace exactly what is already there and then be required to do it again if development comes.

In fact, the upsized line may become a driving force for development since the needed infrastructure is already in place. Some economic development experts will argue that it is better to make the developer pay for the upgrade, and in some cases this makes sense, but sometimes what attracts the developer is the fact that the water needed is already available. This single factor may make the difference between a project moving forward, because it is now affordable for the developer, and one where the cost forces them to choose a different location, maybe in a different community. The payback for the cost of the upgrade will come through revenue generated by taxes and fees garnered from the newly developed property. Forward-thinking analysis coupled with the courage to act can set a community up for success. This is where strategic planning helps.

Strategic planning should be viewed as a tool to assist leaders in critically reviewing their organizations and making sound decisions based on needs as well as leveraging assets for the future. Leaders cannot effectively manage their organization if they are being reactive or running from one crisis event to the next. To create excellence, you must be proactive and deliberate in your decision-making and actions (Haigh, 2010).

SWOT ANALYSIS

Most strategic planners will tell you that the first step in the process is to conduct an analysis of the organization's status to get a clearer picture of the current situation. Typically, this is done by conducting a SWOT Analysis that looks at the organization's strengths, weaknesses, opportunities, and threats (Heim, 2015). I think of the SWOT Analysis as the research behind the final product.

It must also be understood that strategic planning is both a little science and a little art, and the planning process always seems to vary slightly based on the differences of the organizations and the players involved.

I like to teach that the four aspects of the SWOT Analysis are really opposing views of two distinct areas of focus. Strengths and weaknesses look at the internal environment or the situation inside the organization.

These two focus areas often yield a side-by-side comparison of factors related to leadership and management, organizational support, services provided, performance, people, skills, adaptability, training, and processes.

Conversely, opportunities and threats tend to become a similar side-by-side comparison that looks at issues in the external environment or the situation outside of the organization. Factors tend to focus on resident/customer satisfaction, the economy, politics, demographics, environmental concerns, laws and regulations, media, society, culture, and so forth.

The process of conducting a SWOT Analysis requires free and open discussion, a checking of egos, focus, brainstorming, a willingness to push the envelope, and most important of all, visionary leadership. When assisting organizations with writing their strategic plans, I typically ask leadership to select a diverse team of key stakeholders who understand the inner workings of the department. Depending on the organization, these may be officers, members who coordinate and manage specific duties or job tasks, or personnel who have a long and respected tenure with the department. Leaders need to invite their "A-Team" members to participate in this process. This does not mean your "yes team" or your members prone to group think but rather the best of the best who can provide critical thinking and vision. These members need to be able to set aside personal agendas and focus on the big picture of where the organization is and where it needs to go. I usually recommend between 10 and 12 members who bring not only the expertise (science) but the experience and intuition (art) required to effectively plan for the future. In smaller organizations, the number of participants may be less. The key is to have enough members involved to capture a holistic view of the organization but not overwhelm the process (fig. 2–2).

Team members must be willing to speak the truth and professionally debate the various areas being discussed. Leadership must approach this process with an open attitude and be willing to hear and accept constructive criticism. When you end up with a poorly developed strategic plan, you need to look no further than this work group. Selecting this team is the single most important aspect of the planning process. The entire organization must be able to believe that the right team members have participated in the process or the

Figure 2–2. Board members from the trade association Associated Equipment Distributors (AED) discussing their SWOT Analysis.

plan will have no buy-in from the organization's members and will simply be disregarded (Haigh, 2016; Wallace, 2006).

CONDUCTING THE SWOT ANALYSIS

In the early days of doing strategic planning, I would facilitate a group discussion with the members of the planning team whereby we talked about each individual category of the SWOT. As we talked, I would list out the generalized topics discussed in a bullet-point format on an easel chart while making notes about what each bullet-point statement meant. After doing several of these, I repeatedly found this process to be incredibly time-consuming and not always productive. It often led to member grandstanding, invited disagreements and arguments, and offered an opportunity for members to lose focus and to run down various rabbit holes. I found that these meetings were absolutely exhausting for both me and the entire team. By the time I ended one of these sessions my emotional energy was running on empty. I knew that I needed to come up with a different plan.

The model I prefer today is to break the group into four separate work groups. I usually go around the room having them number off to avoid the "I always sit with my buddies" group think scenario. This numbering off system seems to work well in shaking up the players and thereby increasing the chances of obtaining good information because each person can contribute based on their own opinion and not be driven by the opinions of their friends.

Once I have the groups divided, I assign one group to focus on strengths, the next to apply their attention to weaknesses, and the third and fourth to work on opportunities and threats. I set a timer for 10 minutes and tell each of the work groups to discuss and list out in bullet-point fashion the issues they think most pertinent related to their assigned section of the SWOT Analysis. At the 10-minute mark, the groups rotate, with those who worked on strengths moving to weaknesses, weaknesses to opportunities, opportunities to threats, and threats to strengths. I restart the clock, and they get another 10 minutes to work. When the rotating group gets their next assignment, they are to review what is already written and make additions as necessary. This process continues until all four work groups have added their insight to each of the lists.

What I routinely find is that with each subsequent rotation, the time needed to complete the work gets shorter and shorter. Sometimes I even begin reducing the 10-minute timer because the groups simply do not need the time. This occurs because the rotating groups tend to find many of their topics have already been placed on the list by the first couple rotations, and now they are down to adding only one or two additional comments or maybe tweaking what is written to add clarity. In the end, you get the very same information in a much shorter time frame.

As the facilitator, I walk around and listen to the discussion occurring within each group. I sometimes ask clarifying questions, may give some careful direction to keep them focused, or invite some of the quieter members of the group to interject their insight to prevent the loudest and most boisterous from dominating the discussion. Once the groups have completed their work, I generally facilitate a discussion with the entire team to make sure that I understand exactly what is meant by each bullet point.

After completing the SWOT and making sure that I understand what has been identified, I tend to adjourn the meeting and schedule a future time to begin writing goals based on the findings of the SWOT. This adjournment does two critical things:

1. It allows the team to go away and reflect on what they learned from the SWOT process.
2. It gives me a chance to take each bullet point and write a brief narrative statement about its precise meaning.

Over the years I have read numerous strategic plan SWOT Analyses that contain simple lists of bullet-point topics. The problem is that not all readers understand what is meant by each bullet point, especially if you were not part of the planning team. This is especially true for newly elected officials or newly hired employees who are expected, based on the nature of their position, to utilize and implement parts of the finalized strategic plan. I also find that as time goes by, even the team members who worked on the project begin to forget the exact meaning behind each listed item. Therefore, I write the narrative statements.

Once I have the narratives written, I send the draft SWOT Analysis to the participating members asking for feedback. I want to make sure that I accurately have captured the sentiments of the discussion and have written the statements so that they reflect the true picture of the situation. Figures 2–3 and 2–4 are two examples showing parts of a SWOT Analysis from East Dubuque's 2020 plan and the associated narrative statements.

THE USE OF AN OUTSIDE FACILITATOR

I generally believe that most agencies can work through and develop a strategic plan in-house without the need to hire an outside facilitator. Sometimes I get

Strengths	Weaknesses
Membership and Staffing – The Department enjoys a dedicated staff with a strong commitment to serving the community. The members are extremely resourceful and bring diverse backgrounds and marketplace experience. They have a progressive mindset toward use of technology and are willing to try new things in order to find ways to better serve the community. Members are attracted to the department based on the reputation of the organization, and staffing levels are currently high with numbers ranging between 35–40 members. The department has recently been successful in recruiting members who already hold needed certifications, which minimizes training costs and hastens the new member onboarding process. The "bunk room program" has also become an attractive staffing enhancement allowing members to sleep at the fire station, thereby providing faster response coverage. Response times are acceptable and better than many of the area volunteer departments. This is due to the number of members living close to the fire stations and the relatively small footprint of the response area.	**Apparatus** – The current fleet is aging with limited available funds to replace apparatus in the near future. Engine 211-1996 Engine 212-1996 Engine 213-1981 Squad 231-1991 Boat 251-1998 Maintenance costs are increasing, and the reliability of these aging assets is becoming a concern. Engine 213 is at the end of its useful lifespan with Squad 231 not far behind. A tremendous challenge in finding replacement apparatus, new or preowned, is the severe space limitations at both fire stations to house modern-day apparatus.
Department Leadership – The Department is blessed with a proactive and involved group of officers who bring a multitude of marketplace disciplines to the department. The department has enjoyed limited officer turnover, which has served to stabilize the organization. The officers are dedicated and committed to growing and developing the staff. They lead using an open and nondictatorial style where members feel free to share ideas and feel a strong sense of organizational ownership. This leadership style also garners tremendous respect from the community, city staff, elected officials, and other emergency response agencies/professionals.	**Stations** – The current fire stations are not geographically positioned to maximize response times for both responding members and timely response to emergency incidents. Their locations are not hardened and make them susceptible to flooding, train derailments, and damage from falling boulders. Space is limited, and apparatus floors are not large enough to accommodate modern, non-custom-built apparatus. Neither fire station has shower facilities. This prevents members from following best practices for personal decontamination post fire incidents. They also have no gear washer/extractor system to remove carcinogens from turnout gear or dry the gear once washed.

Figure 2–3. East Dubuque's 2020 SWOT Analysis strengths and weaknesses.

asked to serve as a facilitator, and I am happy to do so, but I would coach readers to not be afraid to give strategic planning a try on their own. By doing the project in-house, I believe it brings more personal ownership to the plan, and it tends to be used as intended rather than simply being placed on a bookshelf awaiting the next update.

KNOWING WHEN YOU NEED HELP

There are times, however, when it is advisable to hire an outside facilitator. Depending on the political climate of your organization, sometimes it is easier and safer to have an outsider who can come in, ask the tough questions, probe deep on the "sacred cow" topics, and then leave the meeting with head and neck still attached to shoulders. When you anticipate an internal struggle, it is usually wise to reach out for assistance.

Opportunities	Threats
Ambulance Service - With the addition of the ambulance service, the potential exists to recruit additional members interested in providing EMS services. This also includes service expansion opportunities that could potentially incorporate local nonemergency ambulance transports and evaluate mobile integrated health care/community paramedicine. The department needs to continue the evaluation process for adding a second ambulance to the fleet. This would allow the department to handle multiple patient transports and provide a backup unit when the primary unit is down for maintenance.	**Member Retention and Succession** – Failure to plan for member attrition and replacement, including succession planning. Many current members are not interested in serving in formalized leadership roles/officer positions. This is of great concern regarding overall succession planning for the future of the department. This is compounded by current member aging and health concerns. Additional concerns revolve around a nonresident candidate pool that brings a large percentage of members from surrounding communities rather than from the City of East Dubuque. This has the potential to lengthen response times to emergency incidents as personnel need to travel further to get to the fire stations.
Recruitment – Solicit membership opportunities from area businesses and neighboring communities. Offer specified training to businesses willing to release their employees for emergency response purposes. Develop a specialized recruitment process for members who live within the corporate boundaries of East Dubuque.	**Station** – Facilities are inhibiting growth and department development and service mission. Both facilities are located within known hazard zones and are not designed to accommodate current and future service demands.
Update/Relocation of Fire Stations – Due to challenges with station design, size, and location, significant attention needs to focus on correction of this issue. With either station expansion or new facilities being constructed, the potential exists to add a multipurpose room that can be used for community-based functions as well as large training classes and enhanced public education programs. Identification of possible land acquisition sites need to be a priority in order to address this opportunity.	**Vehicles** – Reliability of existing apparatus due to age and overall capabilities. Included in this list is the lack of an aerial device as well as a second ambulance.
Training – The department is fortunate that the City is supportive of outside training and attendance at conferences even if this includes travel. Work is ongoing to continue enhancements to the training facility. Planned enhancements include area lighting, props, and an additional live burn cell. Other planned enhancement includes a firefighter escape/bailout simulator, and a confined space training system. In addition, the training site provides the opportunity to allow usage of the facility by other fire departments as well as the Fire Service Institute. This shared usage strengthens mutual aid relationships and improve firefighter training within the region.	**Funding** – Availability and acceptance of grants, including line-item funding through the City's budgetary process. Reimbursement for supplies used at emergency incidents through failure of the cost recovery program. Limited grant success due to the East Dubuque Volunteer Fire Department Association not having 501c (3) status. A relatively new concern is the number of instances where EDFD is requested mutual aid for an ambulance response only to be canceled en route. This causes a financial hardship in that the City pays a stipend to responding members with no capability to recover costs associated with the response through ambulance billing. This concern is driving by the challenges neighboring volunteer ambulance services are experiencing with staffing.

Figure 2–4. East Dubuque's 2020 SWOT Analysis opportunities and threats.

Remember, egos often get involved in this type of work, and without an outside player to help everyone stay focused on the overarching goal, the process can unravel and not produce the needed results. Although you may be qualified and capable of managing the process, sometimes it makes sense to let an outsider take the lead. Assess your situation and do not be afraid to give it a try, but also know when help is needed.

ASKING YOUR CONSTITUENTS

Some organizations also like to conduct a second SWOT Analysis with community members to get a feel for how the organization is viewed at large and whether it is meeting the perceived needs of its constituents. In these cases, a facilitator is also very helpful.

For this type of SWOT Analyses meetings, I like to first have someone from senior management provide a basic overview of the organization and the services offered. This overview is often extremely helpful to the public members who have been selected to help with the assessment process. It is amazing how little the general population really knows about what we do and what is required to effectively provide services. Then as the facilitator, I give some general directions and use the same process detailed before to conduct the SWOT Analysis.

WRITING GOALS

Once you have the SWOT Analysis finished (the research aspect of the project), you need to next begin writing goals. The goals section of the plan focuses on what must be done to address areas identified in the SWOT Analysis.

I usually find that many of the topics identified in the four different categories of the SWOT are very similar or interrelated. When viewed holistically, solutions become obvious. The goals section of the plan is where these solutions are brought forward and an implementation plan is designed (Haigh, 2016).

Whenever I think about writing goals, I always think back to my days attending instructor school at the fire academy. The process for writing goals and their associated objectives is the same, whether for a fire training program or a strategic plan. Goals relate to the desired result you want to achieve and are typically broad and can be long term. Objectives, on the other hand, are the specific measurable actions needed to accomplish the goal.

In most cases, after careful analysis of the SWOT, the number of goals gets boiled down to a relatively small list. I usually recommend trying to limit the number to between five and eight. When you think about the numerous objectives that are required to accomplish a single goal, creating a limited number of goals tends to make the plan more realistic for the selected time frame.

GOAL FORMATTING

I like to use a chart for my goals. There is nothing special about this format model other than I find it to be succinct and easy to read. It provides a clarity of the work that needs to be accomplished.

Using the chart format, I list the goal to be accomplished at the top of the chart and then create four side-by-side columns. I title the first column "objective" (or sometimes I use the word "action" or the phrase "action step" based on the group I am working with and what works best to help them understand what is needed). In this column I list out the various objectives required to accomplish the goal. The next column is titled "measure of success," and for each objective, I list out how we will know when the objective has been met. Remember, objectives need to be measurable. As my friend Heather Moore, deputy director of the Illinois Fire Service Institute, always asks, "What does right look like?" You need to describe in this column what right looks like. The third column is labeled "who's responsible," and the fourth column, "target completion date."

The overarching goal of developing a strategic plan is to make necessary changes that affect the organization's operation. With the very busy schedules that fire service leaders juggle, it is easy to push off work on a specific plan area for another day. When this occurs, departments will regularly reach the end of the planning period with no real work accomplished. It is very easy to fall into this trap because we are all busy. The division of work between various individuals or groups based on areas of expertise, along with the establishment of deadlines, becomes beneficial in ensuring that the work gets done (Haigh, 2016). Figure 2–5 shows an example of a goal/objective chart that was taken from the 2020–2022 East Dubuque plan.

PLAN TIME FRAME

Lastly, as it relates to time frame of the plan, I urge organizations to focus on a plan that lasts no more than three years. In the past I have worked with longer plans, but I am finding that the fire service of today is hugely dynamic, and longer-term plans almost always become obsolete before their projected end date.

PROFESSIONAL LOOK OF THE PLAN

The look of the finalized plan is important. Plans that have a professional look and feel tend to not get lost in the generalized office piles of paper and truly become documents that are used to help guild the organization into the future. When putting together a strategic plan, I like to create an attractive cover page that utilizes photos that project pride or help tell a story about the status of the organization. I have used pictures of stations, rigs, personnel, special projects, or a photo collage of various items. The pictures selected need to paint a visual picture of the message contained in the strategic plan. Often this is something

Goal: Expansion or Replacement of Fire Stations

Action	Measure of Success	Who's Responsible	Target Completion Date

Conduct a needs and space analysis of both fire stations. Consideration points:
• Apparatus housing needs divided between two stations:
  o (2) Engines
  o Rescue/Engine
  o Ambulance
  o Reserve Ambulance
  o Command Vehicle
  o Ladder/Aerial
  o Rescue Boat
  o Air Trailer
  o Antique Engine
	Analysis complete	• Command Staff
		• City Leadership
			October 2019

Evaluate physical location of fire stations. • Floodplain considerations • Efficiency of response • Impact of rail lines	Analysis complete	Command Staff	October 2019
RFP to establish cost estimates. • Station 2 Replacement • Station 1 Replacement	RFP developed, sent out for bids, and cost estimates established	Command Staff	December 2019
Evaluate grant funding options	Analysis complete / Potential funding sources identified Prepare plan to submit grant applications	Command Staff	February 2020
Develop and receive approval for a funding plan as part of the FY 2020-2021 budget process.	Funding plan established	• Command Staff • City Leadership	May 2020
New Fire Station No. 2	Station open and operational	Command Staff	May 2021
New Fire Station No. 1	Station open and operational	Command Staff	May 2022

Figure 2–5. A goal/objective chart from the 2020–2022 East Dubuque plan.

that shows how the membership is striving for excellence and working hard to serve the community, but it can also be used to highlight challenges and how the plan will help to address deficiencies.

Also, do not ignore the back cover. I often use this area to print additional photos or display a large logo or a graphic of the organization's tagline. I think of this section as the "final word" or an area that can be used to convey the overarching message you are trying to send in the strategic plan. It is the last page the reader will see and should be used to help them remember what they just read. The lesson here is to not be afraid to be creative and to use the front and back covers as billboards advertising the overall message of the plan. I usually print and laminate my covers or have them professionally printed on hard card stock and then bound. This gives them a finished, professional look that conveys the importance of the document (fig. 2–6).

WHAT'S INSIDE

Although the cover will help to keep the document from getting shelved and ignored, it pales in comparison to the written information contained inside. I like to start my reports with a narrative that provides an overview of the strategic planning process:

- The overarching goal of the plan
- The organization's mission, vision, or values statements
- Names of the members involved in development of the plan
- How long it took to do the work
- Any trainings, presentations or special research projects that were conducted to help the team focus on specific areas

I also try and provide some basic information about how the organization operates. It is important to see the strategic plan as a tool to share information with readers who may not be completely familiar with the agency. To that end, I usually insert a graphic depicting the organizational structure or an organizational workflow plan to help clarify what the reader will be seeing later in the document. Following the overview, I next insert the SWOT Analysis along with the narrative statements associated with each SWOT topic followed by the goal charts.

FINANCIAL PLANNING DOCUMENTS

Two other sections that I like to incorporate into most plans are a Capital Equipment Planning Chart and an Apparatus Replacement Planning Chart.

Chapter 2 Strategic Planning 37

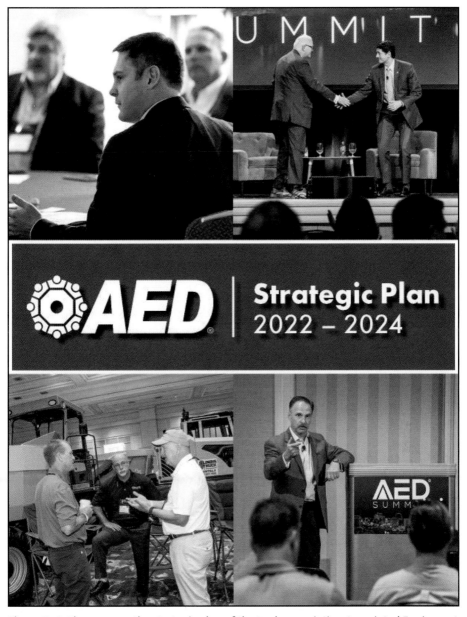

Figure 2–6. The cover on the strategic plan of the trade association Associated Equipment Distributors (AED).

Both documents are explained in detail in chapter 3—Managing Money. I find their inclusion in the strategic plan serves to create a direct linkage between planning and expenditures and works as a great tool to help focus the overall budget process.

As an example, if a department has an upcoming large expenditure such as the purchase of new SCBA somewhere in the next 10 years, and it is identified within the Capital Equipment Planning Chart, the organization may want to develop a goal to create a sinking fund (i.e., savings account) where money can be set aside for this future large-scale purchase. This planning process becomes even more essential when it comes to apparatus replacement simply due to the high cost of replacing these machines.

THE VALUE OF PHOTOS

Due to the diverse group reading our strategic plans (e.g., department members, elected officials, members of the community, other governmental organizations) I often find it helpful to insert photos and a brief explanation of some of the capital and apparatus items listed on the planning charts. Not all readers are going to understand the terms used to describe items, and the photos help tremendously to ensure that the writers and the readers both understand what is being discussed.

As an example, firefighters understand that the abbreviation SCBA is referencing breathing apparatus used for toxic atmospheres or those immediately dangerous to health and safety (IDLH). SCBA is different than SCUBA, with the later referring to breathing apparatus used for underwater work. Both may be listed on the capital planning chart but refer to two very different, yet both expensive, pieces of equipment. A photo can help identify the difference.

Similarly, the term "fire truck," when read by someone outside of the fire service, could mean anything from a pickup truck to a water tender (a vehicle used primarily to haul loads of water from a water source to the fire scene—used when hydrants are not readily available such as in a rural setting). When someone from the fire service hears the term "truck," they think of an apparatus with a hydraulic extendable ladder used to reach elevated heights or apply water from an elevated position. Statements that talk about "purchase of a truck" can easily be confusing, depending on who is reading the document. Photos and brief explanations are incredibly helpful to ensure that both the writers and the readers understand what is being discussed in the plan.

MANAGING THE PLAN

Once the plan is complete and has been adopted by the organization's governing board, periodic accountability and status reviews are a must. Quarterly reviews work well for many organizations, while others prefer semiannual ones. Either way, leadership needs to pull together the management team and the responsible parties on a regular basis to check progress and ensure that the details of the plan are moving forward (Haigh, 2016).

As these reviews are completed, the team will likely find that some goals or objectives need to be modified. This is normal and should be expected. When one digs into the specifics of a situation, whether to correct a deficiency, build a new program, or bolster something that is already in place, we often find roadblocks or things we missed that must be addressed—this is all part of the process (Haigh, 2016).

When conducting these reviews, it is important to not get discouraged if you do not get every objective or goal completed. This is not a failure of the plan. Plans should be dynamic and flexible based on changing conditions. Unexpected situations arise. Sometimes after digging into a topic identified in the plan, you learn information that sends you in a different direction. All of this is okay and should be expected.

BUDGETING AND FINANCES

In addition to the quarterly/semiannual reviews, the plan needs to be used as the guiding document when constructing the annual budget. The budget should reflect the plan. A plan that does not have budget support for the goals and objectives or that waits until the final year of the plan to receive budgetary attention is a plan destined to fail. The bottom line: if you are going to begin using strategic planning as a management tool, it needs to be used to also drive the budget.

CHANGE MANAGEMENT

As much as I believe in the value of strategic planning, we must understand that planning is sometimes (maybe more often than not) met with resistance. Strategic planning by its very nature is about change, and change can be hard. For the leader trying to implement the newly developed plan, this singular issue often becomes the most daunting aspect of the work.

Change, or the resistance to change, is an interesting topic. Change is essential, and all organizations need to transition in order to maintain relevancy in meeting the needs of those they serve. I think that most people understand this basic fact and do not resist change simply out of principle, but rather they resist change based on a fear of personal loss. Authors Ronald Heifetz and Marty Linsky, in their book *Leadership on the Line*, wrote, "Although you may see with clarity and passion a promising future of progress and gain, people will see with equal passion the losses you are asking them to sustain" (Heifetz & Linksy, 2002, p. 12).

Resistance to the plan sometimes comes in open attacks that often become personal as emotions flare. Other times the attacks are behind the scenes as naysayers work to circle their wagons and plan resistance to the change. The key to remember is that in most cases these naysayers are not bad people but rather people who are afraid of the unknown. As a leader, you need to be able to stomach this hostility while staying connected to the people. This connection becomes the driving force in getting people to tackle the problems and the changes identified in the plan. If you allow yourself to disconnect, this disengagement only exacerbates the problem. Heifetz and Linsky remind us, "You appear dangerous to people when you question their values, beliefs, or habits of a lifetime. You place yourself on the line [as a leader] when you tell people what they need to hear rather than what they want to hear" (Heifetz & Linsky, 2002, p. 12). In order to appear less dangerous, it is essential that you remain connected.

Quality strategic plans are stretching and create a level of risk and organizational instability. Plans that are not at least a little edgy probably did not dig deep enough into the various issues and will ultimately prove to be not very valuable. Plan development also needs to consider timing, messaging, and key influencers as you work to implement the plan. Leaders need to be keenly aware that change requires disturbing people, but it must be done at a rate that they can absorb. The lesson here: proceed carefully!

CELEBRATING THE SUCCESS

When the planning period comes to an end, and you have weathered any storms, worked through the issues, and are beginning to think about and develop a successor plan, it is a great idea to provide a close-out presentation so all involved (e.g., department members, elected officials, citizenry) can clearly see the progress made and can celebrate the organization's success. This allows everyone in the organization to see the progress achieved and the plan's impact. This also helps to generate excitement, buy-in, pride, and a sense that the organization is moving in a positive direction. It works to instill the importance of planning in the minds of those who will lead the organization in the future, which helps to build an organizational legacy of excellence.

WORTH THE EFFORT

Strategic planning is challenging, hard work, sometimes a little dangerous, and time consuming. However, it is an outstanding tool in the arsenal of a leader who desires to meet the needs of those they serve by creating organizational excellence and eliminating the reactive management model. The strategic planning tool, when used correctly, will help to ensure success in leading your emergency service organization (Haigh, 2016).

REFERENCES

Bouffard, W. (2011). *Puttin' Cologne on the Ricksaw: A Guide to Dysfunctional Management and the Evil Workplace Environments They Create*. Willima L. Bouffard.

Collins, J. (2001). *Good to Great: Why Some Companies Make the Leap . . . and Others Don't*. New York: HarperCollins Publications, Inc.

Haigh, C. A. (2010, February). *Strategic Planning: Planning the Village's Future* [Presentation]. Village of Hanover Park Board and Elected Officials, Hanover Park, IL.

Haigh, C. A. (2016, August). The Art of Planning. *Fire Engineering*, 67–70.

Heifetz, R., & Linksy, M. (2002). *Leadership on the Line*. Boston: Harvard Business School Press.

Heim, J. M. (2015, January/February). Planning for the Future of the Fire Service. *The Bulletin, Illinois Firefighter's Association*, 18–19.

Lee, T. H., & Duckworth, A. L. (2018, September/October). Organizational Grit. *Harvard Business Review*, 98–105.

Wallace, M. (2006). *Fire Department Strategic Planning: Creating Future Excellence*, 2nd ed. Tulsa, OK: PennWell Corporation.

QUESTIONS FOR FURTHER RESEARCH, THOUGHT, AND DISCUSSION

1. From a strategic planning perspective, answer the following questions related to your own organization:

    a. What are the unique needs of your community, and what services *should* your organization be providing based on these needs?
    _____
    _____
    _____

    b. What tools, equipment, staffing, and programs are required to meet today's service demands? What is missing?
    _____
    _____
    _____

    c. How is the community changing? How will this impact service demand in the future?

    i. Likely changes/service demand changes in 3 years:
    _____
    _____
    _____

    ii. Likely changes/service demand changes in 5 years:
    _____
    _____
    _____

    iii. Likely changes/service demand changes in 10 years:
    _____
    _____
    _____

2. In preparing to conduct a strategic plan for your own organization, list 10 individuals you would select to serve on the planning team and why.

    a. Name: _____ Why: _____
    b. Name: _____ Why: _____
    c. Name: _____ Why: _____

d. Name: _____ Why: _____
e. Name: _____ Why: _____
f. Name: _____ Why: _____
g. Name: _____ Why: _____
h. Name: _____ Why: _____
i. Name: _____ Why: _____
j. Name: _____ Why: _____

3. Based on your personal experience, conduct a SWOT Analysis of your own organization. List five topics under each category:

   a. Strengths:

      i. _____
      ii. _____
      iii. _____
      iv. _____
      v. _____

   b. Weaknesses:

      i. _____
      ii. _____
      iii. _____
      iv. _____
      v. _____

   c. Opportunities:

      i. _____
      ii. _____
      iii. _____
      iv. _____
      v. _____

   d. Threats:

      i. _____
      ii. _____
      iii. _____
      iv. _____
      v. _____

4. Using your SWOT Analysis, write a goal to address an identified issue.

Goal:			
(i.e., what you want to accomplish)			
Objective/action steps	Measure of success	Who's responsible	Target completion date

CHAPTER 3

MANAGING MONEY

Financial management can be scary stuff! When considering whether to pursue a promotional opportunity into the top spot, the single most frightening new responsibility for many fire chief/CEO candidates is financial management. They understand that they will be judged on their ability to develop and manage a budget and that mistakes can be career ending. The problem is that most fire officers have never been trained in financial management, have spent little time trying to figure it out, and often have been left out of the budgeting process by their superiors. They understand that money is important, but they would be perfectly fine in most cases to allow someone else to take care of such things as long as they have the funds needed to operate their organizations effectively.

As an example, I was recently helping a candidate prepare for a fire chief hiring process. As we were discussing the various fire chief/CEO responsibilities he commented, "I just want to do the job, not deal with all that other stuff." The "other stuff" he was referencing was financial management. The problem is that the "other stuff" is a huge part of the job of the fire chief/CEO.

I think most fire chiefs/CEOs feel good about their ability to handle emergencies, manage their staff, and serve as the face of the department. However, the phrase "budget time" strikes fear into the hearts of these battle-hardened fire service veterans and turns their nerves of steel into flimsy rubber bands.

City block on fire—no problem!
A thousand-acre wildland fire—we got this!
Multicasualty plane crash—triage, treat, and transport!

Labor/management meeting—no big deal!

Talking to the city's finance director—ABSOLUTELY TERRIFYING!

Firefighters become good at what we do through training and experience. We spend countless hours attending classes to gain knowledge and learn new skills. We have constructed multimillion-dollar training facilities where we simulate all types of scenarios so that firefighters can be placed in realistic learning environments prior to facing them at a real incident. We have fire officer certification programs that address strategies, tactics, leadership principles, training, and fire prevention, but we spend almost zero time teaching the aspiring fire chief/CEO how to develop and manage a budget. The truth of the matter is that most of us find our way to the top spot having never been trained in even the basics of government finance.

I will acknowledge up front that financial management is not necessarily my thing. I like to do planning and vision casting, but I am not over the top excited about digging into the minutia of each debit and credit detailed on a balance sheet or general ledger. I get excited about what projects are possible through sound money management, but I am not necessarily eager to spend time reviewing financial reports. I enjoy seeing the success that comes after the hard work of planning, strategizing, and being wise with financial resources. But the day-to-day budget management does not rev my motivation engine.

I will also say that what I know about financial management I have either figured out on my own or picked up from a few finance people willing to share their experience and knowledge. I do not have an accounting degree like my son-in-law, and I would say that I am not the sharpest tool in the shed as it relates to math. I am simply a fire chief who realized early on that to be successful, I need to understand how to manage the money.

This chapter is about what I have learned and what has helped me be successful in navigating the world of government finance. It is not a chapter on accounting practices, government financial regulations, or even best practices, it is a down-and-dirty, firefighter-focused overview of some of the principles that I have learned and find helpful in managing the money.

TRUSTED WITH A LITTLE . . . TRUSTED WITH A LOT

I am reminded of the words of Jesus found in Luke 16:10: "Whoever can be trusted with very little can also be trusted with much, and whoever is dishonest with very little will also be dishonest with much."

When I stepped into the volunteer fire chief role at Hampton Fire Rescue in 1991, we had an annual budget of around $60,000. I tried very hard to use every penny wisely and knew that the village board would intensely scrutinize my every purchase. I learned to be a wise shopper making purchases based on best prices, and I also came to appreciate the value of buying preowned equipment and tools. As an example, we designed and ordered a new custom heavy rescue truck that was delivered with reconditioned hydraulic reels, a cascade system, front and rear winches, and an onboard air tool compressor system that were all purchased preowned and sent to the factory for installation. Even the Federal Signal Q2B® siren was repurposed from another unit and placed on this new rig. The monthly publication *The Pennsylvania Fireman* published by the Lancaster County Firemen's Association became a trusted source for departments selling items they had declared surplus (Lancaster County Firemen's Association, n.d.).

I made it known in as wide a circle as possible that we were a department open to equipment donations. Just because it was worn or even broken did not mean that we were not interested. We had a couple of very gifted volunteer mechanics and fabricators who could repair most things, thereby allowing us to get a few more years of life out of a needed tool.

These early experiences now drive my passion to get tools, equipment, and even apparatus into the hands of small organizations who can use the things we as larger organizations are upgrading or switching out. I am constantly on the lookout for equipment that a small department/district can use. I cannot stand seeing tools and equipment taken to the scrapyard if they still have some usable life. This goes for apparatus as well. In the past few years working through an organization called Firefighter Hearts United (http://www.firefighterheartsunited.org/) I have been able to help facilitate getting several rigs, tools, and breathing apparatus to fire departments in Ecuador (fig. 3–1).

I now regularly hear from organizations who are willing to pass along their no longer needed assets, and based on my contacts, there is a good chance that I can find it a new home. I take no money for the work and simply feel compelled to help. T. J. Moore, Hanover Park's director of public works, jokingly refers to my tool and apparatus exchange system as "Haigh's Redneck eBay." That is probably a good title.

From my humble beginning back in Hampton to ultimately working in Hanover Park, I feel like God has blessed me with a true Luke 16:10 career. Hanover Park Fire Department is by no means affluent, but they are not hurting. They are not the highest paid department in the Chicagoland metro market, but they are certainly not the lowest. They have exceptional equipment, and their personnel really want for nothing. They are not doing pancake breakfast fundraisers to buy needed tools, they do not purchase preowned equipment, and they have substantial money in the bank. Prior to my retirement, I

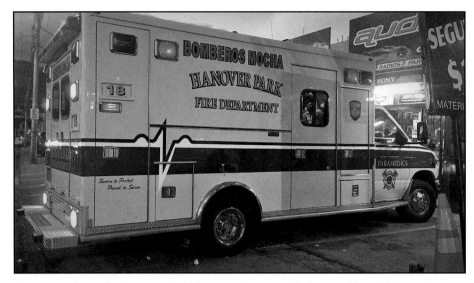

Figure 3–1. A surplus Hanover Park Fire Department ambulance sold and shipped to Mocha, Ecuador. A custom in Ecuador is to leave rigs and tools marked with their original department lettering to honor those who previously used the equipment (photo courtesy of Firefighter Hearts United, LLC).

recognized that this is not the scenario for most fire service agencies across the world, and this realization made me feel extremely blessed.

Through my experiences I have come to realize a couple of key guiding principles related to financial management:

Principle No. 1
If you have been trusted with a little, pay close attention, do not make reckless decisions, and remember that you are a steward of the taxpayers' money. Focus on the fact that the taxpayer has entrusted their hard-earned dollars to you so that you will care for them when they need the services provided by your organization.

Principle No. 2
If you have been trusted with much—follow principle number 1.

ENVELOPE FINANCIAL MANAGEMENT

The basis of my financial management training came as a young boy. Before personal computers, spreadsheets, and online bill payments, my parents taught me budget management through the envelope system. As I worked mowing yards, shoveling snow, and doing other odd jobs, I would take the cash money I was making and divide it up and place it into envelopes based on my expenses. Each envelope represented a budget category or line item (I did not know back then that they were called line items—I just called them envelopes). I had various broad category envelopes and always one for long-term savings.

In the early days there were literally quarters, dimes, and nickels in each envelope. After I got my first real job working as a lifeguard, the envelopes started being filled more with paper money and less with loose change. As my income increased, so did the number of envelopes (i.e., line items). I now needed gas money, money to pay for dates, and money to purchase the things I thought necessary (usually some gadget that I wanted for my role as a volunteer firefighter).

The key principle with this envelope system is that you only spend the money in the designated envelope for what it is supposed to be spent on and you resist the temptation to borrow from other envelopes should you run low. As an example, if my entertainment envelope did not have enough money to buy movie tickets for me and my girlfriend, then we didn't go to the movies. If I wanted new trauma shears for my EMT kit and I did not have the money in the designated envelope, I had to save for a while until I had the money. I was also learning the importance of nonoptional expenses, like gas money for my car. If I ran out of money to purchase gasoline, then I could not go to work, which would then place a serious cramp in my overall cash flow. I always knew that if I really got stuck my parents would help me out, but I was driven to do my absolute best to not ask for help and to be responsible with my financial resources.

This system of envelopes taught me the basics of budgeting that I still use today. I find it remarkable that I had millions of dollars in my budget at the fire department, but I still managed money based on the envelope system. Here is how it worked:

1. We as a department annually figured out how much money we needed to operate (our expenses).
2. We created line items or cost control centers (envelopes).
3. We determined how much money will be available for use over the next year (revenue).
4. We then distributed the revenue into the various line items (envelopes) based on our nonoptional expenses (things that we absolutely needed to pay for—payroll, employee benefits, pension contributions, fuel for the apparatus, insurance, vehicle and station maintenance, replacement tools/equipment, training, public education supplies, etc.).

5. We also maintained line items (envelopes) for long-term savings (e.g., apparatus replacement and big dollar expenses like new rescue tools and breathing apparatus).
6. After all nonoptional expenses were accounted for, if we found that we had more revenue than needed, we began adding dollars into the various line items (envelopes) to purchase tools and equipment that are "nice to haves" rather than "need to haves."

These combined line items, when totaled together, made up our budget. We then carefully watched to make sure that we did not spend more from any single line item than what has been allotted. If we did it correctly, we did not run out of money and we were able to do the things needed (and planned). If we did it wrong and we ran out of money, bad things usually happen.

I realize that this explanation is super simplistic, but organizational budgeting really is not any more complicated than the envelope method I learned as a 10-year-old boy. Do not let the financial industry terms or the hundreds of acronyms throw you. It is nothing more than money in, money out and making sure that you do not spend more than you have available.

UNDERSTANDING YOUR ROLE

Government organizations are controlled by the voters. The electorate votes for candidates to represent their interests as the governing authority of a public body. Public bodies provide services. For the purposes of this example, we will focus on fire/EMS/rescue services. These elected officials are tasked with the responsibility to provide the services demanded by the voters at a level they are willing to pay. This "willingness to pay" is important in that it determines the level of service that a department/district can offer. In some communities, voters are willing to pay for fully staffed emergency organizations capable of arriving on-scene in less than three minutes who can provide the highest levels of emergency services using the latest tools, equipment, and technology to solve their problem. Other communities are not willing to pay for the high cost of a career fire agency and are content to rely on volunteer firefighters and emergency responders. They may not see the need or are simply unwilling or unable to pay for the best apparatus and equipment. They understand the ramifications of having fewer services available and are willing to take the risk.

Elected officials hire or appoint managers (e.g., fire chiefs/CEOs, village/city/county/township managers, and administrators) to provide professional advice on how best to accomplish the demands of the electorate. These demands become the organization's mission. The elected officials then raise funds to pay for accomplishment of the mission. This is usually done through property taxes or fees for service. If the electorate is not pleased with how the elected officials managed the mission, they replace them at the next election. I believe that it is important for the fire chief/CEO to understand this process and their role in the grand scheme of things.

I have watched fire chiefs/CEOs, me included, who advocated, argued, and made ourselves generally crazy about an elected official's (or in proxy, a decision by their village/city manager's) unwillingness to act on a recommendation designed to enhance service levels. Yes, fire chiefs/CEOs should be using fire service best practices and professional standards to develop recommendations and should always be looking for ways to enhance their level of service and professionalism. But we also need to realize that sometimes what we believe is right for our community or district is not what the voters see as the highest priority and they are therefore unwilling to pay more in tax dollars. This is a tough pill to swallow for the passionate and driven fire chief/CEO who only wants the best and is now getting a lesson on willingness to pay.

As the fire chief/CEO you should give the best advice you can and then recognize your place in the vast workings of government. Based on my own experience, the sooner you understand your role, the faster your stress level will return to a manageable level and the better relationship you will have with your elected officials and bosses.

STRATEGIC PLANNING'S ROLE IN FINANCIAL MANAGEMENT

I was once told by a village manager that the budget is the planning tool for the village. I respectfully disagree. I do not believe that the budget is the planning tool. Rather, it is the tool that allows the plan to be accomplished.

Once the elected officials establish the mission of the organization based on the direction they have been given by their electorate, planning needs to begin on how this will be accomplished. The planning aspect comes through the strategic plan. Strategic plans typically cover more than a single budget cycle, which is important in that larger projects often take several years to accomplish. The planning process identifies what needs to be done. Goals are developed as well as a listing of the specific objectives needed to accomplish the goal. Once the goal is accomplished, budget dollars are then used to sustain the goal over the long term.

As an example, a department/district may set a goal to purchase a ladder truck. Over a three-year strategic planning process, they work to obtain funding, write specifications, order, have built, and take delivery of the new apparatus. From start to finish, this process can easily take three years. However, once the new rig arrives and is placed into service, it will require constant maintenance to keep it serviceable. Ladder trucks are expensive to purchase and equally expensive to maintain. If the budget process only includes dollars to make the one-time purchase and does not include annual dollars for maintenance, the department/district will soon find themselves with a nonserviceable million-dollar liability.

To sum up this example:

- The strategic planning goal was to purchase the rig.
- The budget was the tool used to set aside the money to buy it.
- When the rig was purchased and placed into service, the goal as identified in the plan was accomplished.
- Now the budget needs to continually include funding to maintain what was purchased in order to fulfill the mission established by the elected officials and as dictated by the electorate (to provide quality fire protection services).

In other words, the budget was used to accomplish the plan—it is not the plan.

REVENUE: TAXES

A basic tenet of financial management is that expenditures cannot exceed revenue. If you need to increase expenditures in order to support the mission and you find that you do not have enough revenue to cover the expenditures, you must first work to increase the amount of available money. For most governmental agencies, the primary source of revenue is taxes.

Taxes and tax laws are complicated and vary from state to state. Each state can determine what powers it will grant to local governments, including regulations related to taxation. This authority of the state to provide oversight and allow local government only the rights expressly granted to them by the legislature is commonly referred to as "Dillon's Rule" (National League of Cities [NLC], 2016). Some states, however, have adopted into their constitutions or by statute the standard of Home Rule, which, depending on how it is written, allows municipalities/counties/townships expanded rights related to the passage of laws governing themselves. Some of these laws may include local taxation (NLC, 2016).

Property taxes are often used by local government (e.g., municipalities, counties, townships, fire districts, school districts, library districts, park districts) to derive the funds needed to provide services. Property tax is based on the assessed value of a property, including any buildings or structures (known as improvements) located on the property. Depending on applicable state laws, property taxes can be either rate-based or levy-based.

The rate-based system establishes a "tax rate," which is then multiplied against the assessed value of the property. The higher the assessed value, the more tax dollars paid. Under this system, property tax collections increase or decrease as property values fluctuate. To help stabilize revenue, a few states utilize a levy-based system. The levy-based model allows taxes to be assessed based on a specific dollar amount levied each year. If a drop in property value occurs, the tax rate automatically increases in order to collect the same amount of money. Likewise, if property values increase, then the tax rate decreases in order to garner the same dollars. Often tax caps are applied by state law to regulate how much this levy rate can increase annually without first seeking approval from the voters (City of Shoreline, 2016). Based on the tax caps, it is conceivable that a significant reduction in assessed value, even under the levy system, will result in fewer dollars coming into the organization through taxation.

However, this issue of tax caps does not apply evenly to all states and jurisdictions. As an example, the state of Illinois in 1991 enacted the Property Tax Extension Limitation Act (Public Act 87-17) (Illinois Library Association, 2015). Although impacting most taxing bodies, the new law exempted Home Rule entities. Home Rule entities are required to hold a public hearing should they intend to increase their levy by more than 5% from the previous year, but they do not need to seek approval from the voters (Property Tax Code, 1994).

Voter approval on the other hand occurs through a referendum. A referendum is a public question placed on a ballot by a local unit of government (State of Indiana, n.d.). Without passage of a referendum, except in cases such as Illinois's Home Rule exemption, a public body is unable to increase their taxable limit beyond what has been established by law.

Referendums are of such importance that the Illinois Fire Protection District Act actually provides within the statute an illustration of how a question should be asked on the ballot:

> Shall the maximum allowable tax rate for the ... Fire Protection District be increased from ... % to ... % of the value of all taxable property within the District as equalized or assessed by the Department of Revenue?
>
> YES   NO

(Fire Protection District Act, n.d.)

WHEN TAX DOLLARS FALL SHORT

Depending on the assessed value of taxable properties, non-Home Rule organizations can find themselves in a difficult position if they are unable to generate enough revenue through property taxes to effectively operate the department/district. Reliance completely on property tax without the ability to generate revenue in other ways can quickly create a massive financial challenge. Municipalities typically have additional taxing options available (e.g., sales tax, food and beverage tax, hotel/room tax, Home Rule sales tax, income tax sharing) while fire protection districts are often much more limited. Some states allow districts to charge for services or licenses/permits, and some assess fines for fire code violations, while others prevent such fees/charges.

The best advice I can provide to the fire chief/CEO who is trying to learn and understand the complexities of their state's tax laws and allowable fees for service is as follows:

1. Read the general statutes of your state related to taxation, specifically looking to see how they apply to your organization. Also read the statutes related to fire/EMS/rescue agencies and their granted authority within your specific state. When I say read them, I mean actually read them. Do not rely on hearsay or the opinion of others. We have lots of firehouse lawyers who believe that they know the law yet have never attended law school nor have they read the applicable statutes. Most states now have their statutes posted online, and with a quick search you can easily find what you are after.
2. Hire a competent attorney who specializes in municipal or fire district law to provide advice on how tax laws apply to your organization. Also use them to assist with the drafting of ordinances and regulations related to taxation and fees for service. It is not wise to try and go it alone in addressing these type of issues.
3. Begin assessing alternative revenue options in order to address the funding gap between what you receive in taxes and what is needed to effectively operate your organization.

ALTERNATIVE REVENUES

If your organization is granted authority by state statute to generate revenue beyond just property taxes, you need to determine whether this income is a one-time revenue or a sustained revenue.

- One-time revenue: generated and typically used to support a special project. There is no guarantee that it will be renewed or sustained into the future.
- Sustained revenue: annually repeatable and can be counted on for operational budgeting.

ONE-TIME REVENUE

When thinking about one-time revenues, grants and fundraisers are probably the most common. Volunteer departments for years have used creative events to generate revenue required to pay for equipment, training, vehicles, and other operational needs. Anyone who has held membership in a volunteer department/district has most likely been involved in pancake breakfasts, chili suppers, dances, gun raffles, carnivals, bingo/casino nights, parking cars at the county fair, or any number of other events designed to generate revenue. According to the *National Fire Protection Association's—Fire Service Needs Assessment*, fundraising makes up 15% of the budgets of volunteer fire departments protecting communities with populations of 2,500 or fewer residents. There are more than 14,000 such fire departments in the United States (National Volunteer Fire Council, 2020). Although these events can be fun, they can also be time consuming, exhausting, and a drain on volunteers with limited discretionary time.

As a general rule, I include grants as a one-time revenue, although some, such as the Federal Emergency Management Agency (FEMA)—Staffing for Adequate Fire and Emergency Response (SAFER) program can be used to cover on a limited basis the cost of salaries, benefits, and ancillary expenses for firefighters (FEMA, 2020). The key is that the grant is limited. When the grant award comes to an end, the department/district will need to absorb the cost of these new employees into their annual budgets or will be forced to lay them off or attrition them out.

Numerous other grants are available, ranging from the FEMA—Assistance to Firefighters Grant (AFG) to funding opportunities from private entities, businesses, and corporations. Some states also offer grants for small tools or zero interest apparatus loans. Often these are managed through the state's fire marshal's office.

A major difficulty of grant programs is that they are competitive and often challenging to find, and the application processes can be extensive and time consuming. My friend Chief Jeff Bryant of the Amboy (IL) Fire Protection District is the best I have seen in processing and obtaining grants. He has literally brought in millions of dollars to his paid-on-call/part-time organization. However, to become successful he had to become a student of the process and learned both the technique of writing grant applications and then how to manage them once awarded. Applications need to be clear and succinctly written, depicting the need without being long and overly burdensome to read. Departments are often challenged in finding someone who can devote the time needed to both locate the grants and then process the applications.

Corporate sponsorships are another source of funding that was once commonly used to cover the cost of special events but in recent years has become more of a source of regular income for some organizations. This will be an interesting trend to watch as the cost of providing services continues to increase while budget dollars get tighter.

As an example, the Ashburn Volunteer Fire & Rescue Department, part of the Loudoun County (VA) Combined Fire & Rescue System, offers corporate sponsorships in levels of bronze, silver, gold, and platinum with monthly contribution levels ranging from just over $25 to nearly $500 (Ashburn Volunteer Fire & Rescue Department, n.d.).

The Mesa (AZ) Fire and Medical Department has implemented a sponsorship opportunity giving businesses the chance to feature a public safety message along with their branding logo on the side of fire trucks. This move, which one of my North Carolina friends recently described as NASCAR for fire trucks (referencing the sponsorship logos displayed on race cars), raises $25,000 annually. Deputy Chief Forest Smith says related to the message and logos, "We are extremely respectful to the fact that we want to keep a fire truck looking like a fire truck. But at the same time, it gives us that platform to get these important messages out" (M. Moore, 2017) (fig. 3–2).

Another source of income that seems to be an expanding trend is funeral memorial gifts being designated to specific fire departments/districts. More and more people are moving away from sending flowers or wreaths and opting for something more lasting and useful, such as a monetary memorial. These gifts, given in the name of the deceased, are often driven by their personal connection to the organization or because the organization provided valued and appreciated services to the deceased (e.g., compassionate care by the organization's ambulance service during their years of declining health). Similarly, several departments/districts have also established formalized bequest programs where cash, securities, or other properties can be transferred to the organization through a will or trust. These programs are developed, legally reviewed, and advertised as a giving option for individuals doing their final estate planning.

Figure 3–2. The private/public partnership in Mesa, Arizona, was designed to offset budget reductions but has led to a self-sufficient community education program. Private/public partnerships such as this create an opportunity for businesses to invest in a department's mission while supporting and educating the community on health, fire, and life safety initiatives (photo courtesy of Mesa Fire/Medical Department).

A word of caution related to memorial gifts and bequests. Generally, a department/district needs to have a formalized process for charitable funds received on behalf of the organization. Often this is a firefighter's association or a similar 501c3 entity where gifts can be directed. Without this, and especially with municipal departments, the funds often get collected and deposited into the general fund where they quickly are lost in the mix of all other revenue. When this occurs, the memorial or bequest loses its intended impact on the organization. Departments/districts that plan to establish a formalized bequest program need to identify a specific process through which these funds will be collected and utilized.

SUSTAINED REVENUE

A variety of sustained revenue options exist depending on what is allowable by law and what will be supported by both elected officials and their constituents. Billing for responses and patient care has been a mainstay of departments providing ambulance service for years. However, the fire service is seeing more and more cases where this "fee for service" model is extending to other call types as well.

As it relates to ambulance transport billing, this topic, like tax laws, is complicated and is much more than simply sending out an invoice to patients and asking them to remit payment. Ambulance and patient care fees can generally be broken into four distinct billing/payment categories depending on the individual patient's situation:

- Private insurance
- Medicare
- Medicaid
- Self-pay

Each of these categories, with the exception of self-pay, have specific rules that apply as they relate to allowable charges, customary fees per region of the country, and whether you can balance bill one form of insurance against another (i.e., bill private insurance to cover the balance of a transport not paid by Medicare). For ambulance providers doing nonemergency (convalescent) transports in addition to their emergency work, the issue of "medical necessity of need" applies. In this case, before an ambulance provider can get paid for services provided, a physician needs to sign off attesting to the fact that the transport by ambulance was medically necessary. Transports without this signoff, unless self-pay, typically go unpaid.

Likewise, standards also exist related to treat/no transports. A common situation is the diabetic patient who has low blood sugar and ends up being treated with IV dextrose and then, once fully alert and oriented, refuses transport to a medical facility. Also applicable are invalid assist calls such as patients needing help moving from a car to back inside their home or the invalid person sitting on the toilet and simply needing help getting up. All of these situations are ones in which we respond and, in some cases, have the ability to charge and collect fees for our services.

Unless your agency has the ability to support a billing department that is able to stay abreast of the complexities and ever-changing requirements of insurance claims, standards, and collections, I believe it best to contract these services to specialized medical billing companies. Most of these companies will take a percentage of what they collect as payment for managing the billing process and then remit the remainder to the department/district. Numerous examples of ambulance insurance and Medicare fraud exist. Ignorance of billing laws is not a legally defensible excuse; therefore, if you are going to bill and receive funds, you need to make sure that you are following the law. Most fire service organizations today are not large enough to have this billing expertise maintained and managed in-house. I find it better to simply hire the experts and allow them to handle this aspect of revenue generation. Their fee for service is far less than the fines assessed for an error in Medicare billing.

Beyond ambulance billing, many departments have initiated a variety of fees for service. Many of these fees can be processed as insurance claims. Depending on the nature of the incident and the associated charge, the fee may be applicable to a driver's vehicle insurance or a homeowner/business policy. I continue to recommend that if you are processing charges against an insurance claim, the private contractual billing firm will work best for collection of these fees. Some of the more common charges are the following:

- Fire suppression/hazard mitigation
- Vehicle fires (e.g., autos, commercial vehicles, trains, planes)
- Salvage services (e.g., water removal, structural board up, use of salvage covers)
- Vehicle extrication
- Technical rescue (e.g., collapse, trench, confined space, high-angle, water rescue/recovery)
- Hazardous materials response and mitigation

These type of fees for service can be controversial for many communities. Even though in most cases they are allowed as a covered expense by insurance carriers, many elected officials believe that these services are covered by taxation and are therefore unwilling to implement such charges. I am also aware that some states block fire/rescue agencies from billing for services.

Several years ago, I led the Village of Hanover Park through the initiation and implementation process of creating a fee for service model and codifying it into local ordinance. We refer to our program as "cost recovery" (Fire Prevention and Protection § 46-74, n.d.). The argument made, which has been supported by our village board and residents, is an understanding that taxes pay for service availability (i.e., equipped, staffed, trained, and ready to respond). If services are utilized, however, there will be a fee for service. This allows the village to operate at a lower tax levy with the fees for service used to make up the difference in the required budget dollars.

Beyond fees for service and cost recovery programs, other creative sustained revenue options exist. Here are a few to consider:

- *Team of record contracts*: The Occupational Safety and Health Administration (OSHA) standards require industry to maintain detailed emergency response plans as well as specific operational standards for hazard materials, fire brigades, and permitted confined space entries. Some departments have entered contractual relationships with industries to serve as their "team of record" and thereby provide services (standbys, response, and training) in order to meet the OSHA standards (see *Emergency Planning* 29 C.F.R. 1910.38, 29 C.F.R. 1926.35; *Hazardous Waste Operations and Emergency Response* [HAZWOPER] 29 C.F.R. 1910.120(q); *Fire Brigades* 29 C.F.R. 1910.156; and *Permit-Required Confined Spaces* 29 C.F.R. 1910.146(k), 29 C.F.R. 1926.1221). One of the most common is for technical rescue teams to do on-site standbys during confined space entries. Since these entries are scheduled, it allows fire agencies to call back team members and not commit on-duty resources to do this type of work. These team of record contracts can be quite financially lucrative while providing great experience for our special operations team members.

- *Special event standby fees*: Many communities have venues that serve as gathering places for concerts, sporting events, rallies, conferences, and the like. With the vast size of these venues and the number of attendees, many insurance carriers require clients to have EMS services on-site and ready to respond during events. This is an excellent opportunity for departments/districts to serve as the contractor and provide these services. In Hanover Park we have incorporated language into our cost recovery ordinance related to standby services and fees as well as the fact that all standby services must be provided by the department and not an outside private entity:

  > When a fire company (engine, truck, or ambulance) is requested or required to stand by for a nonvillage-sponsored event, the person responsible shall be required to make restitution . . . to the Village of Hanover Park for the costs of such fire company standby. No person shall contract for fire company standby services within the village other than with the village fire department unless the village is unable to provide such services. (Fire Prevention and Protection § 46-74 (2)(e), n.d.)

- *Community rooms*: Taking the lead of volunteer departments, paid fire agencies are beginning to construct large community rooms attached to their facilities that serve as rental venues for birthday parties, bridal and baby showers, anniversary parties, retirement parties, meetings, holiday gatherings, and so forth. These community rooms are often in high demand and provide a sustainable revenue source for organizations that have had the forethought to construct them as part of a fire station building project. One of the great draws to these types of facilities is their location. Since fire stations are constructed throughout a community for quick response times, users of the facilities typically try and rent the room attached to "their" fire station and therefore do not need to travel far for a quality and safe gathering place. An attractive aspect for many is the fact that the facility is attached to a fire station, which becomes a great public relations opportunity for the fire department/district.

- *Infrastructure rental fees*: As technology expands, many telecommunication companies are looking for antenna sites. Since fire stations require antennas for communication equipment, the use of a radio tower or unipoles as a rental site for telecommunication equipment is a lucrative public/private partnership that can generate significant dollars for departments/districts.
- *Training fees*: Several departments/districts have constructed outstanding learning facilities that include classroom space as well as hands-on training props. They also have staff members with expertise and experience that may not be available within smaller agencies. This is a great opportunity to leverage the use of these facilities and your personnel to conduct tuition-funded classes. Depending on how the site is configured and what props are available, training classes may range from full fire academies to officer development programs. Other options might include EMS certification classes, health care provider CPR, CPR/First Aid/Stop the Bleed classes for the general public, and a CPAT testing site. This may also be a great opportunity to partner with private industry to conduct employee training in the OSHA-required emergency response standards.
- *Microdonation programs via e-commerce*: A couple of sustained revenue options I recently learned about are the AmazonSmile® program and the Spare Change® app. The AmazonSmile program donates 0.5% of the price of your Amazon online purchases to the charitable organization of your choice. A registered firefighter's association that qualifies as a 501c3 entity may find this a very worthwhile partnership as the world of online shopping continues to expand. Another interesting option comes from Goodworld (formerly known as Bstow.com) and their Spare Change program, which invites donors to round up on their

purchases and donate the change to the charity of their choice. This program is administered via a smartphone app and can be downloaded for free. Jason Grad, founder of Bstow.com, said, "A lot of these platforms [loose change fundraising programs] leverage mobile technology which is definitely where nonprofit fundraising is headed—mobile donations have grown 45% year-over-year—so if you work for a nonprofit . . . it's definitely a place to start exploring" (Grad, 2016). In order to make these and similar programs successful, marketing by the firefighter's association will need to be done in order for people to know they can make the fire department/district their charitable donation of choice when making purchases through these businesses or using these apps.

FINES AND FD POLICE POWERS

Although the fire service and our personnel have worked hard to brand ourselves as the "good guys," we also need to be cognizant of the fact that some of what we do places us in the unique position to be the enforcement authority for our communities. Depending on the statutes of your state as well as local ordinance, the fire department/district may not only have the power of arrest but also the ability to recover fines for certain violations. As an example, in the Village of Hanover Park, sections 46-35 and 46-74 of the municipal code provide language as follows:

- *Hazardous material incident:* . . . If any person who is liable for the release or substantial threat of release of a hazardous substance fails without sufficient cause to provide removal or remedial action . . . such person may be liable to the village for punitive damages in an amount at least equal to and not more than three times the amount of any costs incurred by the village. . . .
- *Malicious act:* Any person, whose malicious or incendiary act causes an incident resulting in an emergency response, shall be required to make restitution . . . for the costs of that emergency response.
- *Driving under the influence:* Every person found guilty, including an order of supervision or probation, of section 11-501 entitled, "Driving while under the influence of alcohol, other drugs or drugs, intoxicating compound or compounds, or any combination thereof" of the Illinois Vehicle Code, who proximately causes any incident resulting in an appropriate emergency response, shall make restitution pursuant

to the fees . . . for the village's costs of that emergency response. (Fire Prevention and Protection §§ 46-35, 46-74, n.d.)

EXPANDED SERVICES

Back in early 2011 the Hanover Park village manager and I merged fire prevention, the building department, and our health/food sanitarian services into a singular division under the command of the fire department. The focus of this division would be completely public safety–centered with an emphasis on codes, code compliance, permitting, and inspections. The goal was to also create a one stop shop for residents, businesses, and contractors working to navigate the construction process.

Under the plan a new division was added to the fire department called Inspectional Services. This new division became responsible for all aspects of the building permit process from project conception through the issuance of a final certificate of occupancy. All structural, mechanical, plumbing, and electrical plan reviews and inspections are managed by the division. The division also manages all business premises, fire protection systems, health/food sanitation, and commercial property maintenance inspections. To make this happen, the division employs inspectors, permit coordinators, health sanitarians, and licensed architects. The employees of this new division work closely with staff from other village departments to help facilitate code enforcement, zoning compliance, fence and sign inspections, and Freedom of Information Act requests related to buildings and properties.

As part of this new division a multitude of fees for service were implemented. Some of these fees had been charged in the past, while others were new and based on the changed focus of the newly created division:

- Plan reviews (fee is based on a percentage of the total cost of the project)
- Construction permits
- Sign permits (including fees for changes to billboard faces)
- Water/sewer tap on fees
- Annual fire safety/business license inspections
- Quarterly health and sanitation inspections

Fines were also added to ensure compliance for violations such as working without a permit and repeated public safety false alarms. We also initiated a compliance bond program to ensure that work progresses as planned. In certain cases, these compliance bonds can be confiscated due to problems with the construction process.

BUDGET TRACKING AND MANAGEMENT

Once the financial needs are determined, revenue is established, and the budget amount is set for each line item, it is imperative that the fire chief/CEO put in place a system to maintain oversight of expenditures in order to accomplish the goals and objectives funded within the budget. Depending on the complexities and sophistication of the budget, this may look different between different organizations, but a few consistent management practices exist that all agencies should build into their oversight process to help ensure that you end the fiscal period as planned.

The first tool that I use is a report that gives me a quick look at how much I have spent in each line item, how much I have left, and the percentage spent. This percentage-spent calculation gives me a picture of the rate of spending within the 12-month budget cycle. I find it especially helpful for line items that can easily get out of control, such as overtime and vehicle maintenance.

Here is an overview:

> Theoretically, you should be able to take each line item, divide the total dollars by 12 (12-months of the budget year), and get the number you should be spending each month. As an example, if you have $3,000,000 budgeted for your 12 months of payroll, this means that each month you should spend $250,000.

$$\$3,000,000 / 12 = \$250,000$$

> Looking at it from a percentage-spent viewpoint, each month of a 12-month budget should equal 8.33% (100% / 12 = 8.33%). You can then multiply 8.33% by the number of months you are into your budget, and it will give you your expenditure percentage goal. As an example, if you are 3 months into your budget (8.33% x 3 months), you should have theoretically spent 24.99% of a line item's total value.

This percentage tracking model works well for consistent expenditures such as payroll or for monitoring overtime, vehicle fuel costs, utility costs, and so forth. It does not work as well for monitoring accounts that include one-time purchase items. Say you have $4,000 budgeted in a small tools account designated to purchase a new chainsaw and two new nozzles. If you purchase these items three months into the budget year and they have a combined total cost of $3,890, your percentage expended report for that line item will now show 97% spent. This is not a problem in that you have no further planned purchases from this account. The high percentage spent will ultimately balance out as you get further into the budget year. In these cases, I usually place a note in my budget file or check off the expenditure on my "budget detail" sheet so I can remember that this planned expenditure is complete.

Unfortunately, unforeseen expenditures may occur that will overdraw a budget line item. An example would be the need to replace four pair of structural firefighting boots because they became contaminated at a diesel fuel leak. It is almost impossible to predict and budget for these mishaps, so when this occurs you need to either reduce spending in another line item in order to cover the overage or plan to draw from savings (often called reserves) to cover this unforeseen purchase. This is all part of operating a fire department/district and shows the importance of setting aside savings for use as a rainy-day fund.

I utilize a spreadsheet that looks something like table 3–1. The percentage goal used here in this example is based on three months of expenditures.

Second, you need to develop tracking tools designed to do a deeper analysis of potentially problematic spending areas. One that I tracked regularly is employee overtime. In Hanover Park, employee overtime is coded within the payroll system as to why the overtime was worked. Our electronic reporting software has been programmed to accommodate a large array of overtime codes that allow tracking of very specific work scenarios. These scenarios may

Table 3–1. Budget tracking over three months of expenditures.

Line item (envelope)	Amount budgeted ($)	Amount spent ($)	Remainder ($)	Percentage goal (%)	Percentage actual (%)
Salaries—Operations	3,075,086.00	629,673.44	2,445,412.56	24.99	20.48
Overtime—Operations	437,102	54,435.55	382,666.45	24.99	12.45
Small tools—Operations	8,590	2,395	6,195.00	24.99	27.88
Uniforms	27,500	3,426.61	24,073.39	24.99	12.46
Safety/protective equipment	4,550	5645.34	1,095.34	24.99	124.07
Maintenance agreements	46,135.00	24,758.69	21376.31	24.99	53.67

include work related to a special event or standby, the overtime worked by investigators assigned to the regional fire/arson investigation taskforce, or a myriad of other tracking codes we have programmed into the system. The system also allows the codes to be periodically modified in order to track special situations such as our recent experience in managing our COVID-19 response.

Overtime can be both planned and unplanned. Planned overtime is predictable and can be built into the budget based on how often employees have historically worked instructing, investigating, participating in off-duty training, attending meetings, working special events, and so forth. Unplanned overtime is where we can quickly get into trouble. For Hanover Park, this usually occurs in the area of shift coverage. Shift coverage challenge may be driven by a low number of part-time firefighters, by an increased number of shift vacancies due to injury or illness, or by an administration creating vacancies by sending a large number of full-time members to a training class or having them involved in a special project that takes them away from filling their regular company position. When tracking overtime expenditures within categories, I find it helpful to see it visually displayed by a graph or a chart. Most payroll tracking systems can generate various types of graphs and reports to help keep you on track and allow you to see concerns before they become a major problem.

Watching data and tracking trends as a fire chief/CEO is critically important. Trends are predictive of future challenges or needs. They also allow us to see numerically what is occurring within our organizations. This numerical analysis allows us to make timely changes that are not reactive but are based on data. Data helps us make good decisions.

FUTURE EXPENDITURE PLANNING

Capital budget items are those purchases that are expected to last more than a single budget year and generally have a set minimum value. This set minimum value is established by the governing authority. Larger organizations usually have a higher minimum value attached to what they consider capital expenditures compared with smaller organizations. As an example, in Hanover Park they consider any purchase of a single item valued at more than $10,000 a capital expenditure. East Dubuque (IL) Fire Department, where I am a volunteer member, has a much lower threshold. This planning aspect becomes a huge issue when determining revenue related to taxation and alternative revenue needs.

A couple of key aspects of capital asset planning:

1. It is challenging to set capital planning beyond a 10-year time span. The fire service is far too dynamic and 10 years is simply too far out for planning for things other than apparatus or self-contained breathing apparatus (SCBA) replacement, and such plans become at best nothing more than guesses.
2. Since capital assets last more than a single year, you should do your best to predict the overall life expectancy of the asset and then plan for its ultimate replacement.
3. Some capital assets can be placed on a replacement rotation based on life expectancy and NFPA standards. Structural PPE is a good example. If you need to turn over your structural PPE every 10 years—while understanding that some firefighters' gear will have a shorter lifespan than others based on usage—a balanced approach to managing this asset can be planned. As an example, in Hanover Park we know that the average usable life expectancy of structural PPE is from 5 to 7 years. Attempting to provide each firefighter with 2 sets of structural PPE requires that we maintain about 100 sets. Using an average life expectancy of 7 years, we know that we need to purchase around 15 sets annually. We then build this number into our capital asset plan and budget.
4. Large cost maintenance items are acceptable for inclusion in capital planning. Things like interior station painting, roof replacement, carpet/flooring replacement, HVAC, and building tuckpointing are all good examples.
5. When calculating the plan, it is necessary to always build in an inflationary multiplier for each year beyond current. This multiplier varies based on the asset. The best way I have found to determine this multiplier is as follows:
    a. Ask your salesperson how much their product has increased in price each year over the last several years.
    b. Look at the historical increases you are seeing in the actual purchase price.
6. When calculating inflationary multipliers for items that you plan to purchase sometime over the next 10 years, but not right away, you need to find out the cost today as well as the inflationary percentage multiplier and then calculate the inflation each year until your planned purchase time.

7. If you can balance your total capital purchase value so that it is fairly consistent from year to year, it will work better from a revenue planning perspective.
8. In most cases, I recommend handling apparatus replacement separately from this capital asset planning tool. I also personally do SCBA replacement separately as well.

The following spreadsheet is an example of how I typically manage a capital program (fig. 3–3). Note that the total planned expenditures average out to around $250,000 annually over the 10-year plan. By maintaining a consistent capital plan, the fire chief/CEO can calculate and set revenue planning based on the need to bring in around $250,000 annually to maintain capital assets.

SINKING FUNDS/ESCROW ACCOUNTS

Like capital asset planning, departments/districts do well to also plan for future apparatus purchases. The emergency apparatus is the primary tool utilized in providing service to our communities. The challenge with the apparatus is that it is very costly and varies widely in price depending on what type of unit you are purchasing.

As an example, a new command vehicle may cost about $65,000, while an engine could cost as much as $600,000. Ladder trucks and tower ladders are now costing well over a million dollars. These varying numbers make it difficult for most departments/districts (unless you are one of the big guys) to maintain a consistent tax levy when including apparatus purchases into a capital program.

To flatten the variations from year to year so a consistent revenue need can be identified, departments/districts are utilizing a multitude of options to purchase apparatus. Some organizations choose to simply borrow the money like a car loan with a set interest rate paid to the lender. Other departments/

Capital Equipment Replacement Program													
Description	Unit Number	Unit Cost	FY 2022	FY 2023	FY 2024	FY 2025	FY 2026	FY 2027	FY 2028	FY 2029	FY 2030	FY 2031	FY 2032
Firefighter Turnout Gear	15	$3,752	$56,280	$59,094	$62,049	$65,151	$68,409	$71,829	$75,421	$79,192	$83,151	$87,309	$91,674
Tech Rescue Turnout Gear	5	$1,500	$7,500	$7,875	$8,269	$8,682	$9,116	$9,572	$10,051	$10,553	$11,081	$11,635	$12,217
Rescue/Stabilization Tools	1 Set	$47,000						$54,050.0					
Pneumatic Listing Equipment	1 Set	$20,000								$24,000			
Thermal Imaging Cameras	12	$9,372	$18,744	$19,681	$20,665	$21,699	$22,783	$23,923	$25,119	$26,375	$27,693	$29,078	$30,532
Biphasic Cardiac Monitor/Defibrillator/Pace Maker	9	$43,023	$43,023	$45,174	$47,433	$49,805	$52,295	$54,909	$57,655	$60,538			
Extrication Tool Replacement	1 Set	$48,000						$54,308			$59,945	$61,444	
Autopulse CPR Units	8	$16,700		$17,535	$18,412	$19,332	$20,299	$21,314	$22,380	$23,499	$24,674	$25,907	$27,203
Automatic Cardiac Defibrillators	46	$3,000								$72,939			
RAD 57 Monitors	9	$7,000				$63,350	$66,518						
Station 16 Replacment - Feasibility Study		$45,000	$21,750	$45,000									
Station 30 - Kitchen Referb					$25,000								
Station 30 - Floor & Carpet					$30,000								
Station 15 - Training Room Audio/Visual Equipment			$60,900										
Rescue Task Force Equipment	15	$1,450	$21,750						$24,608				$27,842
			$253,197	$204,359	$220,177	$231,186.18	$227,210	$260,205	$263,564	$200,156	$267,989	$215,373	$189,467.28	$253,288

Figure 3–3. A capital equipment replacement program spreadsheet.

districts enter into lease-purchase agreements with manufacturers. Others utilize a turn-in lease program where the organization basically rents the vehicle and then turns it in for a replacement based on mileage.

The model that I prefer is the sinking fund (some call it an escrow fund) (fig. 3–4). In this case an organization makes payments into a restricted savings account, which will be used to purchase apparatus in the future. These payments are typically made monthly.

In order to use this model effectively, the organization needs to determine the future cost of each apparatus. To do this, the department/district needs to identify the estimated life of the vehicle (i.e., how long they plan to own it). They then use a percentage multiplier applied against the original purchase price, compounded annually, to determine the future replacement cost of the vehicle. They then take this number and divide it by the life expectancy in order to determine an annual sinking fund payment. The formula for calculating the future cost of the rig based on compounded interest is as follows:

$$FV = PV(1+i)^n$$

FV = future value (anticipated replacement cost)
PV = present value (original purchase price)
i = inflationary multiplier in years
n = planned service life
Example: $60,000 \times (1 + .025)^8 = $73,104$

If using Excel to calculate the anticipated replacement cost use the future value function:

=FV(H4,F4,0,-E4)
=FV(inflationary multiplier, planned life, 0, - original purchase price)

Note that you should always use the number zero and the original purchase price is preceded by a negative sign.

APPARATUS SINKING FUND
Calendar Year 2022

Fleet ID	Apparatus Identifier	Vehicle Description	Purchase Year	Purchase Price	Planned Service Life in Years	Planned Replacement Year	Inflationary Multiplier	Anticipated Replacement Cost	Sinking Fund Annual Contribution	Monthly Contribution	Current account balance toward Purchase
300	C-1	Command Vehicle	2012	$39,250	8	2022	2.5%	$47,822	$5,978	$498	$47,822
306	B-1	Command Vehicle	2015	$39,958	8	2023	2.5%	$48,685	$6,086	$507	$30,428
361	E-1	Pumper	2009	$408,228	20	2029	5.0%	$1,083,150	$54,158	$4,513	$595,733
362	S-15	Rescue/Pumper	2016	$592,495	20	2036	5.0%	$1,572,066	$78,603	$6,550	$314,413
370	S-15 A	Haz Mat Squad	2014	$269,220	20	2034	5.0%	$714,321	$35,716	$2,976	$214,296
371	T-1	Tower Ladder	2017	$1,155,355	30	2046	5.0%	$4,993,378	$166,446	$13,870	$499,338
374	R-15	Rehab Unit	2015	$258,984	20	2035	5.0%	$687,162	$34,358	$2,863	$171,790
381	M-2	Ambulance	2012	$192,032	10	2023	5.0%	$312,800	$31,280	$2,607	$250,240
382	M-1	Ambulance	2016	$235,587	10	2026	5.0%	$383,746	$38,375	$3,198	$153,499
375	S-15 B	Haz Mat Trailer	2007	$13,254	25	2032	1.0%	$16,997	$680	$57	$8,839

Figure 3–4. Apparatus sinking fund for 2022.

Using this spreadsheet example, fund totals are as follows:

- Required monthly payment: $37,659
- Monthly payment annualized: $451,912
- Current balance of the fund: $2,287,568
- Annual interest income on the account if invested with a 1% return: $22,875.68

One of the biggest challenges in using this model is determining the percentage multiplier. The difficulty comes about based on the number of years that fire apparatus are kept in service. Many departments commonly schedule apparatus to have a 20- to 30-year serviceable life. A great deal can change in technology and standards over 30 years. As an example, when I entered the fire service in 1983, I worked on a 1975 Ford/Alexis pumper. The pumper was built on a two-door F-750 chassis with an extended rear tailboard where members rode to calls. When purchased, this pumper was planned to have a 30-year life with the department. However, in 1990 the NFPA standard changed requiring enclosed cabs and that firefighters stop riding tailboard. This change occurred only 15 years into the life of this 30-year apparatus.

To address the change, the department shortened the life of an almost identical 1965 pumper, which cost $12,424, and purchased a Spartan/Darley pumper as its replacement. The cost of this new pumper was $140,000. No one in 1965 would have predicted an 8.75% annual compounded multiplier would be required to cover the cost of a replacement rig in 30 years, not to mention that a replacement unit would have to be purchased 5 years early due to a standard change (Hampton Fire Rescue, 1997) (figs. 3–5 and 3–6).

Fortunately, we do not regularly see such sweeping changes as moving from the tailboard to an interior seated and belted cab configuration, but changes do occur. I think a more realistic example is the purchase of a tower ladder for Hanover Park. In this example, the village purchased a new tower ladder in 2000 costing $625,000. Department leadership project a frontline estimated life of 15 years. The rig would then move to reserve status for an additional 15 years. The anticipated and planned cost of the replacement tower ladder used a multiplier of 5% which generated $1,237,457 for the future purchase (Parker, 2000). Using the established replacement plan, a new tower ladder was delivered in 2016 at a cost of $1,155,355. We then spent an additional $80,000 to purchase new tools and loose equipment for the rig (T. J. Moore & Haigh, 2016). This is a good example of how sinking fund planning is supposed to work.

Figures 3-5 and 3-6. Hampton Fire Rescue (IL) was forced to shorten the planned 30-year life of this 1965 pumper (top) and replace it early with this 1990 pumper (bottom) based on changes to the NFPA standard. These unforeseen changes make determining the sinking fund percentage multiplier extremely challenging (photos courtesy of Hampton Fire Rescue).

Using the historical cost increases we have seen here in Hanover Park over the last 20 years, they are currently using the annual percentage multipliers as follows:

- Command vehicles, pickup trucks, and light duty inspector vehicles: 2.5%
- Ambulances: 5%
- Pumpers, squads, and tower ladders: 5%
- Trailers: 1%

Depending on future trends, these multipliers may need to be modified, but for today they seem to be working.

Since Hanover Park is a municipal department, they have combined all vehicles into a singular sinking fund. This means that all rolling stock from fire, police, public works, and community and economic development are managed together. This results in an estimated sinking fund value of a little more than $5.4 million. Roughly speaking, they spend about $1 million annually on replacement vehicles (all departments) and currently contribute just over $1.3 million into the fund. With a 1% return on investment, this fund generates around $44,000 annually in interest income. To put this into perspective, that $44,000 equates to the current cost of a new police supervisor's Chevrolet Tahoe. This means that even though the police department is contributing money annually to purchase supervisor vehicles, they could purchase an additional Tahoe each year solely out of interest income. Or said another way, if they miss the percentage multiplier for fire apparatus because of an unanticipated standard change (e.g., moving from tailboard to four-door cabs), they can use interest income to cover the difference. It is this interest income that makes sinking funds my funding option of choice. I would rather collect the interest income myself than pay it to another lending agency. It also gives me room to make an error with a sometimes-nonexact funding process.

As referenced earlier, I also like to use a sinking fund for savings related to replacing SCBA. Hanover Park utilizes a 15-year total replacement life on SCBA units. Based on our last purchase in 2016, they are planning for a replacement of all units in 2031. In order to make that purchase they are contributing $38,012 annually into this fund, which includes a 2.5% annual compounded inflationary multiplier.

A WORD OF CAUTION

If you have been fortunate enough to implement an apparatus/equipment sinking fund, you need to stand ready to protect these funds at all cost!

When elected officials and city/village/township/county managers start to see multimillion-dollar savings accounts they become tempted to either

1. Not fund them
2. Borrow money from them

In either of these cases, you are treading a slippery slope that could cause you to not have the necessary future funds to maintain these essential assets. In times of economic downturn, if you need to make modifications to the fund, it is acceptable to push out the replacement date a year or so if the vehicle's overall condition will allow this to occur. This push can allow a one-time reduction in the annual contribution, but this should not become the norm and should be done with full transparency and an understanding of the impact. Remember—if you do not have dependable fire apparatus, you can't run calls, and fire apparatus are far too expensive to simply "find the money" in the 11th hour after you have failed to plan accordingly.

FINAL WORD

Financial management can be challenging, but it is not impossible. As a fire chief/CEO is it imperative that you muster the courage to try and learn a few basic principles to effectively manage this critical aspect of your organization.

REFERENCES

Ashburn Volunteer Fire & Rescue Department. (n.d.). *Become a Corporate Sponsor.* https://ashburnfirerescue.org/support-us/sponsorship/.

City of Shoreline. (2016, September 26). *Frequenty Asked Questions about 2016 Shoreline Levy Lid Lift.* Shoreline, WA. http://www.shorelinewa.gov/Home/ShowDocument?id=26751.

Federal Emergency Management Agency. (2020). *Staffing for Adequate Fire and Emergency Response.* https://www.fema.gov/staffing-adequate-fire-emergency-response-grants.

Fire Prevention and Protection, Village of Hanover Park Municipal Code § 46 (n.d.). https://library.municode.com/il/hanover_park/codes/code_of_ordinances?nodeId=CH46FIPRPR.

Fire Protection District Act, Ill. Stat. § 70 ILCS 705/14 (n.d.). https://www.ilga.gov/legislation/ilcs/ilcs3.asp?ActID=872&ChapterID=15.

Grad, J. (2w016, December 9). *Cheerful Giving.* Bstow. https://www.cheerfulgiving.com/bstow.

Hampton Fire Rescue. (1997). *A History of the Hampton, IL Fire Rescue—1947–1997*. San Angelo, TX: Tayor Publishing Company.

Illinois Library Association. (2015). *Tax Caps*. https://www.ila.org/advocacy/making-your-case/tax-caps.

Lancaster County Firemen's Association. (n.d.). *The Pennsylvania Fireman*. https://www.lcfa.com/?src=gendocs&ref=About_PAfireman.

Moore, M. (2017, November 28). *Mesa Fire Department Offering Businesses Sponsorship on Fire Trucks*. KPHO Broadcasting. https://www.azfamily.com/archives/mesa-fire-department-offering-businesses-sponsorship-on-fire-trucks/article_4d20b79c-a0f6-5d9f-97db-68f612f5c94a.html.

Moore, T. J., & Haigh, C. A. (2016). *Agenda Item—Purchase of Fire Ladder Truck*. Public Works Fleet Department & Fire Department, Hanover Park, IL.

National League of Cities. (2016, December 13). *Cities 101—Delegation of Power*. https://www.nlc.org/resource/cities-101-delegation-of-power.

National Volunteer Fire Council. (2020, March 31). *Social Distancing Creates Challenges for Volunteer Fire/EMS Fundraising*. https://www.nvfc.org/social-distancing-creates-challenges-for-volunteer-fire-ems-fundraising/.

Parker, C. R. (2000). *Sinking Fund Regarding Fire Department Equipment*. Hanover Park Fire Department, Hanover Park, IL.

Property Tax Code, Ill. Stat. § 35 ILCS 200/Art. 18 Div. 2 Truth in Taxation (1994, September 16). http://www.ilga.gov/legislation/ilcs/ilcs4.asp?DocName=003502000HArt%2E+18+Div%2E+2&ActID=596&ChapterID=0&SeqStart=50900000&SeqEnd=52400000.

State of Indiana. (n.d.). Referendum Information. Department of Local Government Finance. https://www.in.gov/dlgf/referendum-information/.

QUESTIONS FOR FURTHER RESEARCH, THOUGHT, AND DISCUSSION

1. Every organization has base/nondiscretionary expenses that are required to maintain functionality (payroll, insurance, fuel, etc.). Review your organization's budget and provide a listing of these expenses and their budgeted dollar amount.

Expense	Budgeted amount
	$
	$
	$
	$
	$
	$
	$
	$
	$
	$

2. From your own experience, have you ever been faced with a "willingness to pay" situation? What did you do to address the challenge?

   _____
   _____
   _____

3. Applying taxation to your organization:

   a. What authority does your organization have to assess taxes? How does Dillon's Rule apply, or is your organization governed by Home Rule authority?

   _____
   _____
   _____

   b. Do tax caps apply to your taxation authority? If so, how?

   _____
   _____
   _____

c. Based on the most recent reevaluation, what is the total assessed value of all real property within the jurisdictional boundaries of the area you protect, excluding nontaxable property (churches, schools, government buildings, etc.)?

_____
_____
_____

d. How much of this total assessed value is available or allocated for use by your fire/emergency services organization?

_____
_____
_____

4. What is the total annual budget for your organization (operating, capital, nondiscretionary, pension liability, etc.)?

_____

5. If the annual tax dollars generated are less than your annual budget, what other sources of revenue are used to cover expenses?

- Revenue source: _____
- Total dollars generated: _____
- Percentage of this total revenue within budget: _____

- Revenue source: _____
- Total dollars generated: _____
- Percentage of this total revenue within budget: _____

- Revenue source: _____
- Total dollars generated: _____
- Percentage of this total revenue within budget: _____

- Revenue source: _____
- Total dollars generated: _____
- Percentage of this total revenue within budget: _____

- Revenue source: _____
- Total dollars generated: _____
- Percentage of this total revenue within budget: _____

6. There are a multitude of one-time and sustained revenue sources available for use by emergency service organizations. Allow yourself to brainstorm this issue and then provide a list of potential new revenue options. For each, identify whether it would be a one-time revenue or a sustained revenue source.

   a. _____ One-time / Sustained
   b. _____ One-time / Sustained
   c. _____ One-time / Sustained
   d. _____ One-time / Sustained
   e. _____ One-time / Sustained

7. Based on your organization's policies, what is considered a capital expenditure? What is the set minimum value used?
   _____
   _____
   _____

8. Create a 10-year capital expenditure planning document for your organization:

Capital Equipment Replacement Program												
Description	Number of units to be purchased	Cost per unit	Annual inflationary multiplier	Year	Year	Year	Year	Year	Year	Year	Year	Year

9. Create a sinking fund for a piece of apparatus recently purchased by your department.

Apparatus identifier	Vehicle description	Purchase year	Purchase price	Planned service life	Planned replacement year	Annual inflationary multiplier	Anticipated replacement cost	Sinking fund annual contribution

CHAPTER 4

HIRING THE TEAM

Let's start with an exercise....

Grab a listing of your organization's current employees. Depending on the size of the department, this might be a small list where you know each of the employees well. For others, this list may include several hundred names. Focus on the employees whom you interact with the most. In a small volunteer department, you may know all the members and be able to discuss their attributes. In a larger organization, this may be your senior command staff or maybe just the officers. If you are not wearing five bugles yet, think about your individual fire companies or the members of your firehouse. Your goal is to get the list reduced to your key employees. You're looking for the list of the employees that *you* personally depend on to carry out the mission of the department.

Next, mentally walk through the key employees you have identified and think about each one individually: their strengths and weaknesses, their attitude, their commitment to the vision of the organization. Then ask yourself this question related to each one:

> If I knew what I know today and this person was in the hiring process to work here, would I hire them? Yes or no?
>
> If the answer is "yes"—why? If it is "no"—why not?

One of the greatest mistakes we can make related to organizational leadership is hiring the wrong employees. That being said, it is impossible to always get it right. The hard truth is that every fire chief/CEO will make a bad hire occasionally, and no exact formula exists to ensure that every employee will be great. When we do make a bad hire, we need to reflect on the process, assess what happened, and determine ways to minimize the potential of making a similar mistake in the future.

THE VOLUNTEER DILEMMA

Within volunteer departments we often look the other way and hire problem employees because we struggle to get volunteers. We begin to believe that more names on a roster is better and that we can't afford to not take a volunteer—no matter how much of a problem they might be. It's that old question of whether it is better to have a 30-member roster with 15 active members—or a 15-member roster? Likewise, volunteer organizations often have generations of family members serving a community, and an adverse employment decision against one may cause the rest of the family to stop volunteering. This fear (real or not) often paralyzes the fire chief/CEO, causing them to hire/keep problem employees.

THE GOOD EMPLOYEE

It is easy to focus on the challenging or problem employees, but what about our good employees? What do they do that positively impacts the workplace and the job we do on the street? When I ponder this question, I find that it is often easier to identify the good employees than to list the traits and actions that make them good. Being a good employee is far more than simply not being a problem.

I find that good employees are unique, and what they bring to the job varies based on their individual personalities and skill sets. Sometimes they are the employee who quietly goes about their tasks while consistently performing excellent work. Others are out-front leaders who are known, seen, and heard. Their shared commonality is in their passion for excellence and their drive to carry out the overall mission of the organization.

I find that I often spend so much time focused on problem employees that I forget about or neglect the good ones. This unintentional action becomes problematic when my good employees start to feel devalued. I fail to focus on that fact that the good employees are the ones who truly make the organization. Without their contributions, we will fall short in providing the services needed by those who rely on us. It is also important to remember that good employees are not typically high maintenance, and just a little care and support goes a long way in making them feel valued and supported.

SEEING THE BIG PICTURE

Beyond just hiring a good or bad employee, hiring must be viewed as the first step in the succession planning process. With fire service careers spanning 25 years or more and most promotions coming from lower ranks within the local department, the hiring decisions we make today will significantly impact our leadership candidate pool years after we are gone. As a fire chief/CEO, we are very likely getting the opportunity to hire the individual who will be the fire chief/CEO in 25 years. We are hiring them today into an entry-level position and thereby beginning the process for a future mayor, city manager, township supervisor, or district president to appoint them chief. When we begin to look at hiring from this perspective, the importance we place on this process is elevated to a new level (Haigh, 2010).

Helping me to see the big picture, Dr. Brian Crandall, former fire chief and Montana Fire Services Training School deputy director, challenged me to look at hiring numerically. He asked me to evaluate by the numbers how many Hanover Park personnel would someday serve in a leadership position. His argument was that the number is great (probably more than I think) and that based on the numbers, my focus on both the selection and the development of our team members is extremely important.

Taking the challenge, I compared the number of full-time firefighters in Hanover Park against how many of these individuals currently hold an officer's rank. I found that one-third of our employees currently fill formalized leadership roles. This means that for every three employees hired, one of them will ultimately be promoted into an officer's position (lieutenant).

Out of these promoted officers, 45% will receive a second promotion (battalion chief or assistant chief), thereby moving them into a position where they supervise other officers. Of the officers who receive these second promotions, and assuming that the village will fill a vacancy in the fire chief/CEO position from within, there is a 20% chance that one of these officers will be the next fire chief. The likelihood is even higher that someone serving as assistant chief or executive officer will be promoted to fire chief.

Additionally, if I take into account how often they have an acting company officer (firefighter working in a lieutenant's role) covering a vacancy created by a commissioned officer who is off-duty using accrued leave time, I find that about 40% of the time, someone is "acting up." This means that almost half of all employees hired by the department will at some point serve in either a commissioned officer position or an acting officer capacity during their career.

When applying this to new candidate selection, it is important to recognize that nearly half of all employees hired will, at some point in their career, serve in a position where they have direct leadership authority over other personnel. They will be the ones making the decisions on the street and will be directly responsible for the service levels provided to our customers. They will be the face of the department for our residents, businesses, and visitors, and their abilities to perform in times of great customer need will be at the forefront of our service delivery model. Based on these numbers, of every two candidates interviewed, the hiring team needs to ask which one do they believe has the potential to serve in this important role?

I would therefore argue that during a hiring process, we as leadership need to not only look at new candidates related to their capability to perform as quality "blue shirts," but we also need to think about their future "white shirt" (or acting white shirt) potential. Saying it another way, when we hire entry-level firefighters, we need to not only look for those who have the desire and attitude to perform the skills and abilities of a high-functioning firefighter/paramedic but also the characteristics, strengths, and interests that, if encouraged and developed correctly, could equip them to be a future leader.

Admittedly this can be a real challenge when hiring candidates who are in their early to midtwenties and have little to no life experience. Still, I believe we should do our best to look deeper into our candidate's personalities to see what predictive information we can glean. Remember—the military takes candidates from the same age group and turns them into officers who lead soldiers during combat operations. We are not always going to get it right, but I sure think that we should try.

WHO'S ON YOUR HIRING TEAM?

If your goal is to hire quality employees who are a good fit and who will successfully carry out the mission of the organization, you must first begin with selection of the hiring team. In many organizations it is impossible for the fire chief/CEO to be involved in every aspect of the hiring process. It takes time to test, interview, conduct background investigations, and check references. With the workload carried by most fire chiefs today, taking the time to personally be involved in all aspects of hiring is likely impossible. This therefore requires delegation. The hiring team is the group who receives this delegation and is responsible to manage the new candidate hiring process.

When delegating responsibility, the fire chief/CEO needs to ensure that those selected to be part of this team truly understand and take seriously the importance of the assignment they have been given. The direct correlation of the long-term success of the department to the organization's future employees cannot be overemphasized. Hiring team members need to understand the values of the department, the skills and education needed, as well as the soft skills (e.g., communication, teamwork, adaptability, problem-solving, creativity, work ethic, interpersonal skills, time management, attention to detail) that will be required of future employees. They also need to be educated on the legal standards related to hiring practices.

Depending on your organization, the hiring team may include members of the department, representatives from human resources, an outside group of individuals appointed by the elected officials to conduct the hiring process (i.e., fire/police commission), or members of the department's executive team. Some departments hire outside firms to conduct their hiring process. The key aspect in all of this is for the fire chief/CEO to understand that having the right hiring team is a super big deal!

RECRUITMENT: PAID PERSONNEL

Finding the right employees for an organization is far more than simply posting advertisements on job websites and your organization's Facebook page. Potential candidates have choices like never before. Gone are the days where candidates watched their local newspaper for fire department testing announcements and signed up in droves to compete for a place on the two-year hiring list. In years past, our firefighters were hometown folks who grew up, went to high school, and remained in a community. They may have been multigenerational town residents, and getting hired by the local fire department was a sense of great pride to the family name.

Today, our future employees have the world at their fingertips. They investigate the fire departments/districts that spark their interest like a police detective searches for a band of robbers. They check your website, Facebook page, Twitter account, and other social media outlets. They look at pictures of your facilities, apparatus, emergency incidents, training exercises, and employees. They watch YouTube videos posted by bystanders of your people working emergency scenes. They read customer satisfaction reviews, scrutinize your bad press, and check out court cases and lawsuits filed against the organization and its leadership. They review your policies, operational guidelines, and medical protocols. They compare company staffing levels against National Fire Protection Association (NFPA) standards, since they view this as a safety concern related to their own well-being. They look at call volume and shift scheduling. They read your union contracts, compare pay and benefits, and evaluate job enhancement offerings such as tuition reimbursement, leave time policies, and specialty assignments. If they can't find this information online, they consider your department to be behind the times and not transparent. When this occurs, they often move on to a different agency where they can uncover this information and learn about the issues they deem important. It is safe to say that applicants become connoisseurs of fire department employment information and use such to make decisions on where to apply and test.

In the past, organizations often viewed potential applicants with an arrogant posture of "they are coming to us because they want a job." In today's world we need to view applicants as our customers. We need to recognize that we are no longer competing against the community next door for new firefighters but rather against all fire service agencies throughout the United States and Canada. Applicants are willing to move for jobs. They are also deeply interested in how the employer is going to meet their individual expectations and desires.

Additionally, things like the livability of a community and region, residency requirements, school systems for their children, recreational activities, and seasonal climate all play into the overall recruitment of applicants. If we want to attract and hire the best and brightest, we need to learn the importance of marketing our organizations. Failing to understand and spend resources on attracting the best applicants is going to leave our organizations of the future with substandard employees while the rock stars of the fire service wear someone else's uniform and badge.

It is also important to recognize that public safety job recruitment is reportedly down, and many agencies are having significant problems finding interested and qualified candidates. These reduced levels can be attributed to the economy, perception of public safety jobs, a lack of generalized understanding related to the work environment, limited knowledge of public safety pay and benefits, or a myriad of other issues. I do not have a silver bullet fix for any of these issues, but I do think that they highlight the need for serious attention to be focused on new employee recruitment.

RECRUITMENT: VOLUNTEER PERSONNEL

Like a paid department, volunteer departments need to understand the motivational aspects of what attracts someone to become a volunteer/paid-on-call firefighter. Without a keen awareness of the motivational factors associated with being a volunteer, the organization is going to be at a loss on how best to recruit and maintain these essential assets. I would also argue that equally as important as to why they come is why they don't. What are the roadblocks to recruitment, and what can be done to remove them? Also, once you attract a member, why do they leave?

The best advice that I have ever heard on leading volunteer workers comes from a church pastor—another organizational group that depends primarily on a volunteer workforce. He explained that the reason people volunteer is very personal and varies from person to person. The key to managing volunteers is to figure out "what floats their boat" and then build on that unique aspect (Hybels, 2004). In other words, as an organization, once you figure out what floats an individual member's boat, you then need to work on how to keep it from sinking. Your goal at the end of the day should be to have your volunteer workers say to themselves, "I can't believe I GET to do this!"

A key aspect of volunteer member recruitment is an understanding that our current team members are our best (or worst) recruitment tools. How they look and act will define your organization's respect level within the community and whether others want to be part of this group. In order to be a volunteer in today's fast-paced, dual-income, non-community-minded society where weekends are spent with travel sports teams and seemingly never-ending work demands, a member needs to have an inner passion to serve on a volunteer fire department. Because of this juggling of time demands, a member must feel like what they are doing is worthwhile, that it is not a waste of their time, and that it gives them the opportunity to work in a high-quality professional emergency services organization. No volunteer in today's world wants to give their time to a substandard and poorly run organization. They also don't want to be part of an organization filled with drama and interpersonal conflict. Anyone who has ever served on a volunteer department knows that without strong leadership these organizations can easily become a hotbed of drama that serves zero purpose in carrying out the mission of the organization.

ARE VOLUNTEERS EMPLOYEES?

The age-old question that continues to arise when talking with fire service leaders about hiring is whether volunteer firefighters are actually employees. Fortunately, the Social Security Administration provides an answer. Under the common law test (a guide used by the IRS to determine if a worker should be classified as an employee), a volunteer firefighter is considered an employee if they are subject to the will and control (i.e., how they will perform the work) of the person (or entity) for whom services are performed. In other words, if a volunteer firefighter is expected to perform specific skills and tasks under the direction of a supervisor or incident commander or as part of a fire department chain-of-command, regardless as to whether or not they are compensated or paid a stipend, they are considered an employee of the fire department/district where they provide services (Social Security Administration, 2012).

This clarification is important as departments begin to develop and implement hiring standards that are reflective of the job duties provided by volunteer/paid-on-call members. Professionalism and quality of service is not based on a paycheck but rather on the expertise and abilities of the provider. Regardless of whether you are hiring a paid firefighter or a volunteer firefighter, you need to carefully scrutinize your applicants to ensure that you are not only getting quality staff members but that you are also working to protect the public.

THE APPLICATION

The written application is your first formal interaction with those interested in employment with your organization. Some departments utilize paper applications, while others have candidates complete the process electronically. I personally prefer an electronic process. It allows me to utilize searchable fields to categorize candidates based on education, certifications, veteran status, and so forth. This searchability is especially helpful in understanding your overall candidate profile and the strength of your recruitment process, especially if you are looking to reach a specific candidate pool. Regardless of method, the questions asked and the information obtained, either electronically or via a paper, should be the same.

Some organizations have implemented an application fee. This fee is not to generate revenue but rather to weed out nonserious candidates. It is helpful in regions of the country where there are numerous paid departments and candidates do what I like to call "shotgun testing." In this scenario, candidates submit applications to an untold number of departments hoping to just get their foot in the fire service door with plans to later move to their dream department. This is okay if you do not mind spending taxpayer dollars to educate new firefighters who will ultimately utilize the training and experience provided to get a job with another organization.

As an example, in the Chicago metro area where I worked, there are more than 200 paid fire departments all basically recruiting from the same candidate pool. I would routinely see candidates jump between two, three, or more departments before they ultimately land where they want to be. From each department that employs them, they do everything possible to obtain the training and certifications that will make them marketable to another agency.

In most cases this training is provided on the backs of the taxpayers expending the money in hopes of preparing the employee for service back to their own community—not someone else's. Implementing an application fee is often helpful in minimizing this challenge.

A word of caution though: some communities have received significant blowback on charging to apply for a job. It is felt that this fee has the potential to have a detrimental impact on some socioeconomic groups who are looking to better themselves but can't afford the price of admission to a quality government job. I would recommend carefully considering how such a policy might be viewed within a particular community.

INFORMATION ON THE APPLICATION

The application process is designed to obtain personal information about the candidate. There is specific information that should be obtained while making sure that the questions asked do not violate their civil rights. This can be somewhat tricky, so you need to proceed cautiously. These standards change from time to time, and new or more restrictive requirements are added. It is always a good idea to have your hiring packet checked periodically by an attorney skilled in labor law to ensure that you are compliant with the most current and relevant standards. Spending a few dollars up front is far cheaper than the cost of defending a legal action for discriminatory hiring practices.

As an example, you can ask a candidate if they possess a valid driver's license. You can also ask them if they meet the age standards established for the hiring of firefighters (e.g., Illinois General Statute 65 ILCS 5/10-2.1-6.3 requires entry-level firefighter candidates to be between the ages of 21 and 35; Illinois Municipal Code, 2019). You cannot as part of the initial application, however, ask for a copy of this driver's license. The driver's license contains both their date of birth as well as a photo of the candidate. In the initial steps of the process, it could be argued that a candidate was cut from the testing based on their age, gender, or race, which was discovered by seeing the photo and the information contained on the driver's license. As candidates proceed through the application and testing process, you can later request a copy of their driver's license. You will ultimately need it to confirm whether they meet the essential functions of the job as defined in the job description (i.e., ability to operate a motor vehicle). You just can't require it up front when they submit their initial application packet.

Tables 4–1 and 4–2 provide information commonly requested on an employment application as well as questions that are considered out of bounds and not appropriate.

Table 4–1. Commonly requested application information.

Category	Specific information per category
Contact information	Name (including maiden/other names you may have used in the past)
	Mailing address
	Telephone numbers (home and cell)
	Email address
Ability to work in United States	Are you legally able to work in the United States? (Yes/No)
Work history	Job history (usually past three to four jobs)
	Reasons for leaving prior job(s)
	Permission to contact current employer
Educational background (provide copies of diplomas, licenses, and certifications)	Highest education grade you completed
	Formalized education (e.g., college, trade schools)
	Professional certifications/licenses (e.g., EMT, paramedic, registered nurse, teacher license, firefighter certifications)
Military experience	Branch (Army, Navy, Air Force, Marines, Coast Guard)
	Military occupational specialty code (MOS or AFSC)
	Honorably discharged (Yes/No/Other)
Essential functions	Are you able to perform the essential functions of the job for which you are applying with or without reasonable accommodations? (Yes/No)
Language proficiency	In which languages are you proficient?
	Ability to speak
	Ability to read
	Ability to write
Volunteer work	Specific or relatable to this position
References	Provide a list of references (three to five) that can speak to your character, work history, or why you would be a good candidate for this position. References must include a mailing address, phone number, and email.

Table 4–2. Prohibited application information

Prohibited application information
Sex, gender identity, sexual orientation
Age
Genetic information
Height and weight
Race, color, or national origin
Religion
Citizenship
Disabilities and medical history
Pregnancy status
Marital status or number of children
Criminal history (Federal law does not prohibit an employer from asking about arrest and conviction records, yet many states limit the use of criminal records in employment decisions. Check state-specific statutes regarding this issue.)
Military reserve questions ("How often are you deployed for military reserve training?" Military status is federally protected, and employers are not allowed to make employment decisions based on an applicant's current or future military service.)

Candidates should sign and date their applications (electronic signatures are fine). As part of this signature process, you should have them attest to the accuracy of the information provided and authorize you to conduct background investigations. Hanover Park utilizes the following statement:

> By my signature I certify that all statements I have made on my application are true and correct and I hereby authorize the Village of Hanover Park to check the records of various Police and Sheriff's Departments, the Illinois State Police, the Federal Bureau of Investigation, the Department of Defense, past employers and other agencies as the Village deems appropriate to determine the accuracy of the information contained in this application.
>
> I further agree to allow the Village to conduct a check of my credit under the Fair Credit Reporting Act as well as the U.S. Office of Inspector General related to excluded individuals (LEIE) due to conviction from Medicare or Medicaid fraud. I agree that my full legal name, any aliases, maiden, or names used will be provided to the Village along with my date

of birth, social security number or green card for purposes of conducting this background investigation upon the extension of a conditional job offer.

If employed, I agree to comply with the Personnel Rules and Regulations of the Village of Hanover Park and those of the Village's Fire Department. As a Hanover Park employee, I give the Village of Hanover Park permission to use any photographs or videos made of me during my service without obligation or compensation to me including photos and videos taken during the application and testing process.

I hereby certify that the information listed herein is true and agree and understand that any false statements contained herein may be cause for rejection of this application or termination of employment. Failure to authorize any or all aspects of a background investigation will also be grounds for rejection of this application and my exclusion from the testing and hiring process.

NEW CANDIDATE ORIENTATION

Many departments/districts hold a new candidate orientation session prior to the beginning of formalized testing. I believe that this is a fantastic opportunity for candidates to gain some insight into your organization as a potential future employer. I regularly tell candidates that even though all fire service agencies typically provide similar core services, the personality and organizational values of departments vary. I stress that they need to be a good fit for the department and that the department needs to be a good fit them. I also stress that a 25-year or longer career in a fire department you dislike is a very long time and should be avoided at all cost. Therefore, they need to choose carefully where they decide to work. The new candidate orientation gives them an opportunity to learn about the department/district before they waste their time, and yours, going through the testing process for a job that really will not be a good fit.

As the fire chief/CEO who is supervising your new candidate orientation presentation, here are a couple of questions to help assess the message you are sending to applicants (intentionally or unintentionally):

- When do you hold the orientation? I have historically done mine in conjunction with the written exam. This helps focus the candidates mentally on the position and the testing process. It also works to minimize the number of times candidates are required to attend testing and interview sessions. This becomes important for candidates who are traveling from out of the area to compete for a position on the hiring list. This acknowledgment of the logistics and the money associated with making trips for this purpose sends a positive message to your candidates.

I recently watched an Arizona department use Facebook Live to conduct their orientation, which was held several weeks prior to their testing process. This process allows candidates to make a determination as to whether the agency is a good fit, and the delayed time between the presentation and the test gives candidates the opportunity to correct any minor deficiencies before they start the testing process.

- Do your staff members participating in the event make the candidates feel comfortable and present an inviting atmosphere? Do they answer questions in a nonthreatening way? Do these staff members include male, female, and minority firefighters? Are there various tenures of firefighters represented? Does your staff in general, both in their disposition and how they interact with candidates, work to ease or build tension?
- Who is in attendance from senior staff? How do *they* handle questions? Do they project by their demeanor, presence, and interactions the importance of the hiring process? Do they represent the best of the department/district?
- The fire chief/CEO is not always able to attend these orientations due to other commitments, so who represents you when you are not there?
- Who are your presenters? How are they dressed? I am a stickler for having my presenters dressed in Class A uniform. I think that this shows respect for both the organization as well as the candidates and sends the message that the process of hiring our future employees is taken seriously and is of utmost importance. Are your presenters excited, well prepared, knowledgeable, and visibly proud of their department?
- What information is presented, and what is the overall focus of the information? During new candidate orientations we typically only focus on the "sexy stuff" and dismiss many of the not-so-glamorous aspects of the job. We use photos and videos of fire apparatus responding lights and sirens and firefighters doing exciting work like stretching hose, throwing ladders, working a fire line at a wildland incident, or flowing water from one of our million-dollar ladder trucks. We sometimes show extrications, crews using their technical rescue skills, and pictures of training scenarios and recruit academy

videos—all good stuff to get their heart rates up and to develop an excitement for the job.

What we typically minimize, however, is the fact that most of our calls are EMS related. Rarely will you hear someone say, "It's about the medicine, and you need to have a passion for excellence in treating sick and injured people." We don't talk about detox calls, smelly patients, psych evaluations, and the importance of the firefighter's job related to our role in providing human services. We also don't talk about the frequency of the non-fire-focused responses and the skills needed to address these challenging situations. We don't say that in most communities we are the first line of defense when someone doesn't know who else to call and that we are full-service emergency response agencies who occasionally go to a fire—regardless of what the name on the side of the rig says.

Likewise, we don't show photos of firefighters mopping floors, swirling toilets, and hanging around the station simply waiting for the bells to sound.

I often wonder during these orientation presentations if we are doing our candidates a disservice by painting an incomplete and often unrealistic picture of the fire service job. Being a firefighter is a GREAT job! One of the best I can imagine! But do we tell them about the "real" job of a firefighter? I think in most cases we don't.

PHYSICAL ABILITY TEST

Today, almost all paid departments and a great many volunteer departments require new candidates to undergo a physical ability exam. In years past, departments often created their own testing algorithm that focused on what they believed to be important in the testing process. Most of these exams had little to no science behind the process and focused primarily on physical strength and stamina. Many of the testing skill stations were irrelevant and fixated more on washing out candidates than testing their abilities and fitness levels. Most did not have existing firefighters complete the test to validate their relevance or to authenticate the time standards. Often these tests created a disparate impact on female and minority candidates and became a hotbed for lawsuits and legally challenged tests.

In addition to concerns related to the validity of the exam was the issue with the number of candidates injured while taking the tests. In today's litigious employment law environment, it is hard for a fire department/district to defend a candidate's injury, even with waivers, when there is no validation of the testing components.

As I think back to 1988 and the physical fitness test I took to come on the job as a full-time firefighter in Rock Island (IL), it had little to no real relevance to the tasks I would be required to do in the field. I was tested on the following:

- Timed mile-and-a-half run
- Push-up test
- Sit-up test
- Balance beam test (i.e., carrying a 50' rolled section of 2½" hose while walking a 12' balance beam without stepping off)
- Chin-up/pull-up test
- Pulling a charged 2½" hoseline 100' along a level concrete pad
- Ladder climb (i.e., climbing an unsupported aerial ladder extended to 70' at a 65°–70° angle)

By sharing Rock Island's test, I mean no disrespect to them because they were simply doing a similar test to what everybody else was doing. Rock Island is a great fire department, and I am honored to have served within their ranks.

Fortunately, today we have the standardized Candidate Physical Ability Test (CPAT) that was designed in partnership between the International Association of Fire Fighters and the International Association of Fire Chiefs. The CPAT is a timed and validated test that measures how candidates handle eight separate physical tasks or functions, designed to mirror tasks that firefighters would do on the job. The entire test must be completed in a maximum time of 10 minutes and 20 seconds. Testing components are the following:

1. Stair climb
2. Hose drag
3. Equipment carry
4. Ladder raise and extension
5. Forcible entry
6. Search
7. Rescue
8. Ceiling breach and pull

Based on the concerns associated with physical ability testing, including validation of the testing process, I believe it wise for fire departments to stop conducting their own testing and simply require candidates to present a valid CPAT completion certificate. There are numerous CPAT testing sites located across the United States and Canada. Candidates register and complete the exam outside of the testing process for the local department and simply provide proof of test completion. In my opinion, this is a real step forward and has virtually eliminated the concerns associated with this portion of the new firefighter hiring process.

WRITTEN TESTING

Written exams for entry-level firefighters are more than a general aptitude test. Test development is a science unto itself, and these exams are designed to assess specific knowledge of basic criteria found necessary for successful firefighters. Test writers work to ensure validity and reliability and, in many cases, have vetted questions and exams to ensure that their wording and turns of phrase do not send a message that is confusing or produces a disparate impact. The goal is to have a test that assesses specific criteria and that is fair and equitable for all candidates. The test also needs to produce consistent results across testing groups.

Some of the areas commonly assessed through written exams are the following:

- Reading comprehension
- Observation and memory
- Mathematical reasoning
- Mechanical reasoning and understanding of mechanical systems
- Spatial orientation
- Situational judgment

Some of these exams also contain a human relations aspect used to predict candidate job success based on emotional triggers presented in the exam questions. One firm that I am familiar with utilizes a video-based test where candidates watch simulations of common job-specific situations and then answer test questions about what they just saw. Based on how they answer the questions, predictions are made about the candidate in areas such as the following:

- Abrasiveness
- Creating tension

- Authoritarianism
- Inconsideration
- Low tolerance
- Passivity
- Self-focus
- Work avoidance
- Supervisor relations

Testing and assessment firms also exist that will custom-write candidate exams for your specific agency. Since departments are vastly different and their needs vary related to entry-level candidates, it often makes sense to utilize a test that will assess candidate abilities related to the specific needs of the department/district.

As an example, when I served as the fire chief for the King (NC) Fire Department, I needed to hire firefighters who were already experienced and who could not only perform the skills of a firefighter but who could also provide direction and leadership to our volunteers. These new firefighters needed to hit the ground running, and due to our low staffing levels, direct supervision and hand-holding of these new candidates was very limited. This is a vastly different situation than a department who can hire new firefighters, send them to an academy training program, and then spend the next year assessing and developing them through a structured field training officer program.

These two different examples demonstrate the need for variations in new candidate entrance exams. This is where the custom-developed tests truly excel. Firms that do this type of work look at the specific needs of a department and then construct tests that contain valid measurements of the knowledge, skills, and abilities required in candidates for that specific agency. Custom exams are more expensive but well worth the price when you consider the risks associated with making a bad hire.

INTERVIEW: KNOWING WHAT TO ASK

Techniques and formats for interviewing public safety candidates are numerous and are as different as the candidates themselves. Everyone seems to have a favorite technique, question, or process. More than anything, however, the interview should be driven by the needs of your organization.

John Sullivan, a professor at San Francisco State and author of *1000 Ways to Recruit Top Talent*, recommends looking at your top performers when you begin the process of developing interview techniques and questions. What makes them good members of your team? What are some of their common traits? How are they resourceful? What did they do prior to working at your organization that has been successful? He says that the answers to these questions will assist you in the development of selection criteria, which will guide you in building relevant questions (quoted in Knight, 2015).

I think it is also important to look at your organization's values. Each of us has an internal set of personal values which are a reflection of our needs, desires, and what we care about most in life. Values identify who we are and what we hold precious and can be thought of as our internal decision-making reference book. Our personal character is based on our own personal set of values.

Since organizations are made up of individuals, organizations also develop a system of values that are reflective of the people who work there. These collective/shared values become the foundation of your organization and can be used as a basis for selecting new employees.

In Hanover Park we looked at this question of values several years ago right after I came to the department as the outside fire chief. Before I started hiring a bunch of new candidates based on what I thought was important, I wanted to understand the values of the folks who had worked there for years. What did they see in candidates who came and stayed and who fit well within the organization as a whole? Through this analysis we came up with what we call our 80/20 Rule for Hiring New Team Members (table 4–3).

Why so little focus on education and certifications? The way we see it, the fire department is really in the business of training. We do this often, and we do it well. If we hire someone who does not have the education needed, we can fix that. But if we hire someone whose heart does not align with our organizational values, we can't change their heart. Selecting an individual with drastically different values than those of the organization usually results in an employee who does not fit well into the overall culture.

Using your organization's values as a foundational guide, you can begin developing interview questions that work to assess how the candidate would fit into your specific department/district. Questions should be open ended and nonleading (table 4–4).

INTERVIEW: THE PROCESS

Departments must think strategically about the message they send during their interview process. If your process turns candidates off, they will simply move on to other opportunities (Knight, 2015). Like your new candidate orientation, you need to think carefully and strategically about how you conduct interviews.

Table 4–3. Rules for hiring new team members.

HANOVER PARK FIRE DEPARTMENT RULES FOR HIRING NEW TEAM MEMBERS	
Focus: 80% of assessment criteria	**Focus:** 20% of assessment criteria
The candidate's heart	What they know
What is their	What do they bring to the table?
• Motivation? • Passion? • Area of greatest pride? • Level of compassion? • Ability to think and reason? • Commitment to a team? • Fortitude? • Love for the job?	• Education? • Certification? • Prior experience?
Getting these two focus areas reversed normally yields a poor employee fit.	

One of the biggest challenges today is the time needed to conduct and participate in interviews. This challenge exists for both the potential employee as well as the employer. Most fire service agencies continue to rely solely on face-to-face interviews, which are hugely time intensive. Private businesses, on the other hand, have for years used phone calls and video conferencing as part of their interview process. Cody Horton, Walmart's former director of college and diversity recruiting, noted, "People who've grown up using webcams and Skype are very comfortable with this [technology]." These electronic media platforms work well as a preliminary step in whittling down a pool of candidates before having them come in for an in-person interview (quoted in Panancy, 2015).

The use of remote (phone or video) interviews becomes important as we see more and more of our candidates who do not live in the area and would be required to travel multiple times for testing and interviews. Traveling requires them to spend thousands of dollars testing for a position in which they are competing against hundreds of other candidates. The cost-benefit analysis of this situation may simply be too great for the candidate. By using technology to minimize travel issues, we will likely find that we have opened our testing to a wider and more global candidate pool.

Table 4–4. New candidate interview questions.

HANOVER PARK FIRE DEPARTMENT
NEW CANDIDATE INTERVIEW
QUESTIONS

The candidate's heart	What they know

• What interests you about this position?
  ○ What do you know about us?
  ○ What did you do to prepare for this interview?
• What personal achievement are you most proud of?
• What does public service mean to you?
• What role in a team do you prefer?
• How do you handle job duties that you don't necessarily like to do?
• How do you overcome a mistake?
• If you didn't have to work, what would you do?
• What sets you apart, and why would that make you a great Hanover Park firefighter?
	• How do you feel your education and experience has prepared you for this position?
	• What are your future goals?
	  ○ One-year
	  ○ Three-year
	  ○ How are you going to work to achieve them?
	• What is your expectation from this position?

• What is the one question that you expected us to ask but that we didn't?

Along with the issue of timing challenges is the issue of candidate stress. Rebecca Knight, in her article titled "How to Conduct an Effective Job Interview" published in the *Harvard Business Review*, wrote, "Candidates find job interviews stressful because of the unknowns. What will my interviewer be like? What kinds of questions will he ask? How can I squeeze this meeting into my workday? And of course: What shall I wear?" (Knight, 2015).

When stressed, people generally don't perform well. Making them feel more comfortable during the process increases the chance of a more productive interview. This is even true for candidates seeking a job in a high-stress environment, such as the fire service. A good practice is to give the candidates a heads-up in advance of the topics you will be discussing along with a general overview of dress code expectations. I learned in a very real way several years ago that candidate stress levels can be problematic. Fortunately, I had a fellow interviewer who was right on target and caught what I was missing. As leaders we need to be cognizant of this issue and take steps to minimize the stress in order to get to the heart of our candidates.

In 2007, my then training officer, Battalion Chief Jeanine Ames, and I were conducting interviews for new part-time firefighters. I think we also had someone from our village's human resource department with us as well (that is our normal practice). We were well into another long day of interviews, and I was in many ways mentally disengaged. I was tired of hearing the same questions answered the same way by candidates who were simply looking for a fire service job opportunity with no real interest in becoming long-term Hanover Park employees. I was drained, my nerves were on edge, I was thinking of the work piling up on my desk, and I just wanted the day to be done.

The candidate we currently had sitting across the table from us was a polite young man who was well groomed and dressed in a suit and tie. But what I noticed right away was that his hands were not those of someone accustomed to wearing a suit but rather what I would describe as "working hands." They were calloused, toughened, and appeared physically strong with a faint hint of grease staining along the edge of the nails. He was extremely nervous, more nervous than most candidates, and he tried his best to answer our questions, but he was really struggling. He stumbled and stammered through each one as he seemingly tried to give us the textbook answer. I didn't dislike this candidate, but rather I felt bad that he was struggling. My general opinion, however, was pretty neutral, and after listening to his anxiety-driven answers, I was simply ready to cut him from the process and move on.

As we concluded the interview, Chief Ames asked him if there was anything else he would like to add or would like us to know. As I recall, he apologized for not doing a good job with the interview and said that he wished he could do it again so that he could tell us how much he really wanted to be a firefighter. Chief Ames then did something unprecedented. Without seeking my opinion or even giving me a heads-up, she asked if he would like to go outside, take a moment to compose himself, and then come back in and do the interview again. He looked stunned and was clearly taken aback, but he took her up on her offer. She obviously saw something that I was missing.

After a few minutes, Nicholas Ballestra walked back into the room and was a completely different candidate. He was calm, well-spoken, and displayed a passion and motivation for the job, including a commitment to a team-focused mentality, that was far beyond his years. He was exactly what I talk about when discussing the kind of employee I want to hire. He had that "something" that said we should bring him in the door as part of our part-time "farm team" and work to build him into a long-term, full-time employee. At his core he seemed to have white shirt potential with blue shirt skills. I almost let him walk away! If it wasn't for Chief Ames and her willingness to go with her gut feeling, we would have lost the chance to hire this great candidate (fig. 4–1).

Needless to say—we hired him. We sent him to the fire academy and the department's rookie school training program and sponsored him through paramedic school. We had some exceptional team members come alongside him who worked hard to help him grow and develop. Then just a few years later he tested for a full-time position and delivered one of the strongest, most heartfelt, and transparent interviews that I have ever seen. It was incredible! Shortly thereafter it was my absolute honor to approve his job offer for a full-time position and to watch as our mayor delivered the oath of office, officially moving him into his new position.

Today, most shift days Nick is the assigned driver of Squad 15. This company is our department's sole specialized company staffed by elite members trained in hazardous materials management and vehicle and machinery rescue, as well as the technical rescue disciplines of trench, collapse, confined space, and high angle rescue. Nick is a master welder and fabricator and has a mind for mechanical reasoning. He understands how things come apart, how they go back together, and how the various parts and pieces all fit together to make something work. As a member of an elite company where much of the work focuses on issues of mechanical reasoning, he is the perfect employee to have responding to someone trapped or caught in a piece of machinery. His unique gifts when applied through his job are a huge asset to the Village of Hanover Park.

Additionally, he now also serves as one of our field training officers (FTOs). As he recently told me, "I have learned a lot and I hope that I can pass those things on to the probationary firefighters that I am able to teach." Hands down, Nick is a great employee, and I hope someday he will take a promotional exam, because I am sure that he will also make a great white shirt officer.

Here is what I learned from this situation:

- Be cautious of overly stressing candidates during the interview process—you might miss the good ones.
- Surround yourself with people who see things you don't.
- Don't be such a control freak that your team members are unwilling (or afraid) to do something out of the ordinary without first getting your approval.

Chapter 4 Hiring the Team **101**

Figure 4–1. Firefighter/paramedic Nicholas Ballestra of Squad 15 working a structure fire in Hanover Park (photo courtesy of Hanover Park Fire Department).

INTERVIEW TEAM

Rebecca Knight, in her *Harvard Business Review* article, wrote, "When making any big decision, it's important to seek counsel from others, so invite a few trusted colleagues to help you interview. . . . You want to have multiple checks to make sure you hire the right person" (Knight, 2015).

In my experience a team of three and certainly no more than four works well for conducting interviews. This number provides for divergence of opinions while not overwhelming and adding undue stress to the candidate. In Hanover Park the process of employee interviews is handled by our training division. I spent years personally teaching and developing the interview skills of our current training officer and his group of training coordinators. I have now stepped back and allow them to manage the interview and hiring process. They are also joined by a member of the village's HR department. One of the great things about using this team is that the coordinators are, in most cases, not officers but peers of the candidates. They are also union members who provide a unique and appreciated perspective to their view of who we should hire.

It is important to remember that as the fire chief/CEO, someday you will leave the organization, and those who step up behind you need to understand how to do the tasks of leading the department/district. Hiring is one of these essential tasks. Giving your team the opportunity to make critical decisions while you are still here is part of the learning process. Don't be afraid to delegate . . . it is rewarding to see what they can do.

PREFERENCE POINTS

Preference points are added to a candidate's score based on select criteria established from past work and educational experience. Preference points in hiring are established by state statute in some cases, while others are established based on the authority of the governing body of the department/district.

Veterans' preference is the most common application of preference points. These points are designed to show appreciation for a candidate's service to our country. They are also designed to give candidates a leg up when competing against applicants who did not serve and therefore had the ability to attend college, complete vocational training, or gain other experiences that would make them attractive to potential employers. Veterans' preference points in essence work to level the playing field for candidates returning home from duty and looking for employment. To be eligible for veterans' preference points, the candidate must be honorably discharged. Some states also award additional points for veterans who are disabled due to their military service.

In Illinois the Firefighter Hiring Act (Public Act 097-0251) governs preference points. Candidates are awarded points for experience as well as education and even participation in fire cadet programs. Minnesota has a similar preference point system based on training/education and experience (Minnesota State Fire Marshal, 1992/2003/2004/2007/2012).

Another example are the points made available by the city of Chicago to immediate family members of sworn police and uniformed fire department personnel who died in the line of duty. These points are also extended to immediate family members of individuals who served on active duty in the armed forces and who died in the line of duty in a combat zone (City of Chicago, n.d.). I'm not sure how many other departments/districts have a similar policy.

As the fire chief/CEO, before certifying a hiring list, it is imperative that you evaluate the standards applicable to your state or locality related to preference points. Failing to award points appropriately can significantly impact candidates and will add to the likelihood of your hiring list being legally challenged.

ELIGIBILITY LIST

Once all aspects of pre-offer testing are complete and scored with preference points awarded, departments/districts routinely produce a hiring list of eligible candidates. Eligibility on these lists is valid for a set period, often two years, and all hires will be made from candidates on this list. Some departments will hire candidates in rank order based on test scores, while others may use different selection criteria based on their specific needs.

Although there are variations in departmental procedures related to order of testing, almost all departments utilize a written exam, a physical ability test, and an interview. Some departments interview and score all candidates before placing those that pass on the eligibility list. Others place candidates on the list based solely on their written score and then conduct interviews at the time of hire. This latter process is a common practice for large departments that have several hundred eligible candidates and interviewing all of them would simply be impossible. If they need to hire 25 candidates to fill an academy class, they may interview the top 50 scoring candidates and then select from that group. They will then follow the same process for additional groups hired during the life of the list.

POST-OFFER TESTING

Typically, once a candidate is pulled from the list and offered a position of employment, the job offer is conditional depending on the candidate passing some additional testing criteria. If the candidate fails any of these post-offer conditional standards, the job offer is withdrawn, and the organization moves on to the next candidate.

Typical post-offer conditions are the following:

- Satisfactory reference checks
- Verification of candidate's degrees and professional certifications and licenses
- Background check and criminal history
- Fitness for duty medical exam

- Psychological exam
- Drug and alcohol screening
- Proof of eligibility to work in the United States (Thompson & Seidel, 2019)

The reason some of these tests are conducted post-offer is that they may include protected health information that by law can only be evaluated after an offer of employment has been made (Thompson & Seidel, 2019).

In 2009 I conducted a survey of 36 Chicagoland fire departments to determine what types of testing and assessment procedures they use and at what point in the hiring process they are conducted. I was looking specifically to learn what testing components are conducted and when. In other words, once a candidate is pulled from the hiring list and extended a conditional official offer of employment, what happens next? The results of my survey are shown in figure 4–2 (Haigh, 2010).

When you get a candidate into post-offer testing, this is where the pressure of making the right decisions becomes intense. Candidates may be qualified based on education and training and present themselves very well during the interview process. They may be able to achieve great scores on the written and

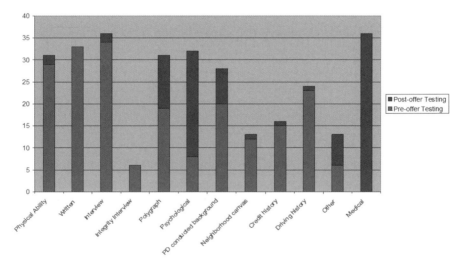

Figure 4–2. Survey results of types of testing and when they are conducted during the hiring process.

physical ability exams, but it is in the post-offer aspects of the process where you either hire a good employee or prevent a long-term mistake.

Once you begin digging into their history and start checking references, you find that many have significant concerns ranging across being poor performing employees, employer theft, illegal drug usage, poor driving records, DUI charges, and extensive credit problems (Haigh, 2009). Although some of the candidate background checks will be completed pre-offer, many of the disqualifying concerns are found in the post-offer phase.

When I studied this issue of disqualifying backgrounds, in Hanover Park we were seeing what I thought was an unusually high number of candidate washouts. At that time, we were losing about half of all candidates due to a problem discovered during post-offer testing. What I wanted to know was whether this was typical or if we had problems in another area of our hiring process—maybe recruitment? To gain some perspective on this issue, I surveyed the group of Chicagoland departments/districts and found that the collective average post-offer washout rate was around 40%. This percentage has now changed, and we are currently seeing job offers being rescinded for about 80% of the candidates based on something found during post-offer testing.

To help work through the hiring process while keeping background concerns at the forefront, I like the checklist/flowchart used by Amboy (IL) Fire Protection District as part of their employee hiring and onboarding process (fig. 4–3).

Common background checks are as follows:

- Criminal history
- Driving record
- Credit check: Remember, employees will have unfettered access to the personal belongings of our customers. It is a good idea to know if they are in a position where they are financially stressed and an easy theft of something during a structure fire might be very tempting. It also goes to their sense of integrity in making promises and then failing to pay back what they have been loaned.
- U.S. Inspector General—List of Excluded Individuals/Entities (LEIE): Look to see if the candidate has been convicted of Medicare or Medicaid fraud. This is important if your service bills for ambulance transport services.
- Sex offender registry
- Fingerprints

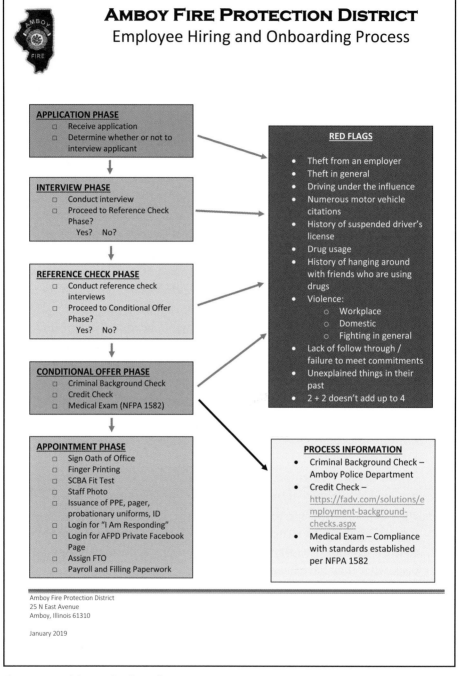

Figure 4–3. Hiring and onboarding process.

- Law enforcement interview, neighborhood canvas, check of employment personnel files: Many departments/districts get assistance from their law enforcement partners in conducting background checks. Law enforcement officers are experts at conducting interviews, asking questions, and following up on inconsistencies. Having a new candidate meet with a law enforcement officer to provide information on their background is very helpful. Based on what is found during these interviews, many law enforcement agencies will talk to neighbors of the candidate and will visit past employers to review personnel files.
- Social media: Some organizations have started conducting reviews of a candidate's social media usage as part of the background investigation. Although this is tempting, it can be a slippery slope and, in some cases, has been prohibited by law. As an example, Illinois has enacted a statute outlining prohibited inquiries of online activities by employers during background investigations. The statute basically says that an employer cannot require or coerce a prospective employee into providing their password or unlocking a social media account so it can be viewed by the employer (Right to Privacy in the Workplace Act, 2012). Similarly, information obtained during a review of social media sources available in the public domain, such as Facebook, can be problematic in that the employer may find information related to race, ethnicity, age, sexual orientation, political affiliation, and other prohibited information. Having this information could prove to be problematic in a discrimination suit based on failure to hire an employee. In other words, it is best to avoid social media, since the risks can easily outweigh the benefits.

FINGERPRINTS

Although hiring rules for public safety employees vary by state, a relatively new law passed in Illinois related to fingerprints is, in my opinion, a helpful step forward:

> A classifiable set of the fingerprints of every person who is offered employment as a certificated member of an affected fire department whether with or without compensation, shall be furnished to the Illinois Department of State Police and to the Federal Bureau of Investigation by the [hiring] commission. (Firefighter Hiring Act, 2011, § 10-1-7.1 (i))

Notice that the statute applies to both paid and volunteer firefighters ("with or without compensation"). I think that over the years many volunteer departments have not done an adequate job in conducting background investigations. I know that when I served as the volunteer chief of Hampton (IL) Fire Rescue back in the early 1990s, I did very little related to background checks and never had anyone fingerprinted. I would say that this was simply naïve on my part. Based on my years of experience, and the weird and surprising stuff I have found, I would now never not conduct a thorough background check on a candidate.

To drive this issue of background checks home, consider for a moment a volunteer department who hires a sexual predator because you failed to identify them during your background investigation. If your department does EMS, think of the free access they would have to patients of all ages in various stages of distress. It is the responsibility of the fire chief/CEO to protect those you serve by not hiring employees who have the potential to take advantage of their trusted position.

REFERENCE CHECKS

I am not ready to recommend that organizations stop doing reference checks, but I am finding them more and more useless in uncovering much information about the candidates. When calling past employers, many organizations have established "neutral reference" policies where they will only confirm the dates of employment that a candidate worked for them. Even with signed waivers from the candidate allowing the department/district to conduct background investigations, many are squeamish about saying anything negative.

When calling personal references supplied by the candidate, they almost always are individuals who give glowing reports of the applicant's abilities. Few candidates are secure enough to provide names of individuals who will provide unfettered truth about the candidate. Occasionally you will get something useful—but not often.

When making these calls, try and ask the references if they can provide names of other individuals who might be able to speak about the candidate. Sometimes you get nothing, but you may get lucky and get a couple of additional names to help expand your search. Occasionally you will find something helpful through one of these contacts.

If a candidate has been a member of another fire department, I will typically call the fire chief. More often than not, this is where I will get the truth. Most fire chiefs/CEOs have been in the same situation numerous times and understand the concerns of hiring a bad employee. Although they may desperately want to give their problem employee away, they usually won't "pass the trash" and allow the candidate to negatively impact another department/district. I appreciate this professional courtesy between individuals who understand the weight of the fifth bugle.

POLYGRAPH

Polygraph investigations have been a mainstay of preemployment post-offer testing for many years. Polygraph machines detect changes in a person's physiologic response when they are not being truthful. It relies on the body's sympathetic nervous system, our fight-or-flight response, producing an increase in heart rate, blood pressure, respiration, and perspiration. The polygraph machine is designed to detect those changes. The theory is that a person who is lying will have an emotional triggering reaction that causes a sympathetic nervous system response, which can be documented and tracked by the machine (Hart, 2020).

Prior to being hooked to the machine, candidates are asked a series of questions known as "pretest admissions." Candidates are instructed to tell the truth related to these questions, no matter how bad, since they will be asked similar questions once connected to the polygraph instrument. The pretest admissions help to diminish emotional reactions. Information gleaned through both pretest admissions and the actual testing process are used to make hiring decisions.

Polygraphers tell us that today's candidates are doing a great deal of research on the overall polygraph testing process in order to find ways to defeat it. Their goal is to figure out ways to not disclose problems related to theft, arson, and drugs and alcohol usage. They are using online chats and discussion forums with other candidates who have been through the testing process in order to learn the types of questions asked. Some candidates have even failed several polygraph exams with different practitioners and are using the experiences to develop ways to try and defeat the testing process through breathing control techniques and practice in answering the questions without allowing a physiologic response.

Minimization is also a common tactic being used to mislead testers. As an example, a candidate may admit to using illegal drugs 2 to 3 times. The actual truth of the admission is that they have used drugs 20 to 30 times. Another example comes from a question about the number of times over the last three years when they have drunk six or more alcoholic beverages in one sitting. They give the answer "a few times." Their statement of "a few times" may actually mean 80 to 100 times. Their goal of this minimization is to provide a truthful answer while not divulging their full drug and alcohol usage.

Fortunately, polygraph practitioners are taking active steps to minimize the candidate's abilities to manipulate the testing.

Some organizations conduct polygraph testing prior to post-offer testing. Use caution in this area in that the testing may by subject to the Americans with Disabilities Act (ADA) if asking questions related to alcohol or drug usage. Answers to these lines of questioning may indicate a dependence issue that could be considered protected health information and therefore undiscoverable prior to an official job offer.

PSYCHOLOGICAL

Public safety psychological testing is conducted by a clinical psychologist specifically trained to evaluate fire service personnel and equipped with an understanding of fire service culture and psychology. They utilize testing instruments specifically designed for public safety applicants that evaluate aptitude and personality. Due to the psychological health aspect of this testing, the American with Disabilities Act of 1990 requires that all preemployment psychological evaluations be conducted post-offer (Steiner, 2017).

Psychologists use several different testing instruments to assess the intellectual abilities of candidates, focusing on their knowledge of words, sentence construction, information retrieval, and quantitative skills. They also conduct clinical interviews to evaluate interpersonal skills, physical health/psychological history, education, work history/work attitude, orientation toward public service, assertiveness, composure, responsiveness to authority, coping skills, and judgment.

Following this evaluation, the psychologist provides a detailed written report to the potential employer. The firm used here in Hanover Park for many years, Stephen A. Laser Associates, rates the candidate in each of the listed categories as follows:

- Strong
- Acceptable
- Marginal
- Poor

The psychologist then provides a recommendation on whether the candidate should be hired based on their professional assessment.

Hanover Park has used Laser's firm for more than 20 years. This group of psychologists knows and understands the culture of the agency as well as the overall organizational expectations required to be a successful Hanover Park firefighter. Their written recommendation indicates their professional judgment of whether a candidate will or will not be a good fit for the department. Leadership has come to so value their analysis based on the long relationship and their track record that a candidate who passes all other aspects but is not recommended by Dr. Laser and his team will immediately have their offer rescinded and will not be hired.

FITNESS FOR DUTY MEDICAL EVALUATIONS

The job of firefighting is both mentally and physically demanding. With more than half of all annual line-of-duty deaths having an underlying medical problem as their cause, it is critically important to ensure that new candidates are medically fit for the arduous tasks of the job.

Prior to a firefighter being hired, they should undergo a fitness for duty medical evaluation. *NFPA 1582: Comprehensive Occupational Medical Program for Fire Departments* has been developed to assist departments/districts by providing best practice information on the evaluation of candidates. This standard does not supersede applicable federal, state, and local laws and regulations and is designed solely for the purpose of providing guidance to medical professionals and department/district leadership. Unless the authority having jurisdiction (AHJ) adopts this standard by reference, including the specific edition, the standard provides no mandate for regulation. This is same for all NFPA standards.

That being said, a lot of really smart people, some of whom are dear friends, have worked very hard to create and maintain this standard as an applicable guidebook to assist in determining firefighter fitness for duty. When sending a firefighter to a physician for evaluation, it is best to find someone familiar with the standard, or at a minimum, you should provide the physician with the standard so they are equipped to correctly evaluate a candidate based on the physical demands of the job.

Likewise, it is important to determine the essential functions of your department and provide the physician with this list. This list should take into account that some firefighting functions and tasks vary based on department size, level of urbanization, and equipment utilized (NFPA, 2017). Also, as part of the fitness for duty medical exam, the physician needs to screen candidates for prohibited drugs.

Candidates should not be appointed unless they pass all aspects of their fitness for duty medical exam and drug screening.

LASTING LEGACY

Hiring is a complicated, time-consuming, and dynamic process. It is imperative that a fire chief/CEO do their absolute best to get it right and to hire staff that will serve the community with distinction and professionalism. But even with the best of intentions, sometimes you make a misstep and hire a problem employee. We all do it, and anyone who tells you that they have never made a bad hire is either not telling the truth or is disconnected from the reality of their organization. The key is not the mistake but how you respond when it is determined that you hired the wrong candidate.

Lastly, if you want to set up a future fire chief/CEO to fail, hire a bunch of problem employees and leave them for the future fire chief/CEO to deal with. The first step in leaving a lasting legacy is hiring the right people (Haigh, 2010).

REFERENCES

City of Chicago. (n.d.). *Line of Duty Preference*. Human Resources, Chicago, IL. https://www.chicago.gov/city/en/depts/dhr/supp_info/line_of_duty_preference.html.

Firefighter Hiring Act, Ill. Stat. Pub. Act 097-0251 (2011, August 4).

Haigh, C. A. (2009). *Disqualifying Backgrounds for Fire Department Employees*. Emmitsburg, MD: Executive Fire Officer Applied Research, National Emergency Training Center, Learning Resource Center.

Haigh, C. A. (2010, August). Legacy of Leadership: Hiring the Right People and Developing Them for the Future. *Firehouse*.

Hart, C. L. (2020, January 14). Do Lie Detector Tests Really Work? *Psychology Today*. https://www.psychologytoday.com/us/blog/the-nature-deception/202001/do-lie-detector-tests-really-work.

Hybels, B. (2004). *The Volunteer Revolution: Unleasing the Power of Everyone*. Grand Rapids, MI: Zondervan.

Illinois Municipal Code, Ill. Stat. § 65 ILCS 5/10-2.1-6.3 Original Appointments; Full-Time Fire Department (2019).

Knight, R. (2015, January 23). How to Conduct an Effective Job Interview. *Harvard Business Review*. https://hbr.org/2015/01/how-to-conduct-an-effective-job-interview.

Minnesota State Fire Marshal. (1992 & rev. 2003, 2004, 2007, 2012). *100 Point Firefighter Employment/Selection Process*. St. Paul, MN: Minnesota Public Safety.

National Fire Protection Association. (2017). *NFPA 1582: Comprehensive Occupational Medical Program for Fire Departments*. Quincy, MA: NFPA.

Panacy, P. (2015). *Why Companies Should Use Online Interviewing Tools like Skype or Google Hangout*. Blue Ribbon Personnel Services, MJT Enterprises, Inc. https://blueribbonpersonnel.com/blog/why-companies-should-use-online-interviewing-tools-like-skype-or-google-hangout.

Right to Privacy in the Workplace Act, Ill. Stat. Pub. Act 97-0875 (2012, August 1).

Social Security Administration. (2012, June 12). *Employment Status of Volunteer Firefighters.* https://secure.ssa.gov/poms.nsf/lnx/0302101260.

Steiner, C. (2017, January 24). Ethics for Psychologists: Pre-Employment Evaluations for Police and Public Safety. *The National Psychologist.* https://nationalpsychologist.com/2017/01/ethics-for-psychologists-pre-employment-evaluations-for-police-and-public-safety/103601.html.

Thompson, J., & Seidel, M. (2019, May 14). *What Is a Conditional Letter of Employment?* Chron. https://smallbusiness.chron.com/conditional-letter-employment-42585.html.

QUESTIONS FOR FURTHER RESEARCH, THOUGHT, AND DISCUSSION

1. Think about the good employees in your organization. What makes them good employees? What traits and actions do they exhibit?

   a. Traits of a good employee:

      i. _____
      ii. _____
      iii. _____
      iv. _____
      v. _____
      vi. _____
      vii. _____
      viii. _____
      ix. _____
      x. _____

   b. What can be learned from the traits/examples of these employees that can be used to guide future recruitment and new employee selection?

      i. _____
      ii. _____
      iii. _____
      iv. _____
      v. _____

2. Evaluate your own organization. What roadblocks exist related to employee recruitment? What can be done to address these challenges? What role will you personally play in bringing about change?

   _____
   _____
   _____
   _____
   _____

3. List the top five reasons that members/employees leave your organization.

   a. _____
   b. _____
   c. _____
   d. _____
   e. _____

4. If you were to revamp your current recruitment process with the goal of reaching the greatest number of qualified candidates, what would the new process look like?

   _____
   _____
   _____
   _____

5. Write five open-ended interview questions for use by your organization to assess candidates for possible employment.

   a. _____
   b. _____
   c. _____
   d. _____
   e. _____

6. Research the general statues and ordinances applicable to public safety hiring within your state or locality related to preference points. Provide a general overview of how these standards should be applied to your organization's testing and hiring practices.

   _____
   _____
   _____
   _____

7. Provide an overview of your organization's post-offer testing process, including applicable general statues and ordinances.

   _____
   _____
   _____
   _____

8. Based on the standards of your organization, list five generalized background disqualifiers related to new candidate hiring.
   a. _____
   b. _____
   c. _____
   d. _____
   e. _____

CHAPTER 5

SUCCESSION PLANNING

"Of course, I have time to meet," I told her. "What time works best for you? I assume you will be coming from the school—so later in the day or early evening? Whatever you need I will make work." I clicked off, wrote the appointment in my calendar, and took a few minutes to think about the conversation.

Wanda Roberson wanted to meet and talk about her son Steven. Wanda was a wife, daughter-in-law, and now mother of King, NC, volunteer firefighters. Wanda's father-in-law, Bill, had retired from the department, and her husband, Steve, and their son Steven were both active volunteer firefighters. Their family had a long and dedicated history with the community, including her own accomplishments as a staff member of the county school system. I was super pleased that she had reached out.

Steven had come up through our Explorer Post/Junior Firefighter Program and was now beginning to contemplate his future. He was a talented young man who had received good marks in high school, was a standout athlete, and had served as "chief" of our Explorer Post. He was someone who just instinctively seemed to understand how to be a firefighter, and people followed him based on his natural leadership abilities.

Wanda and I met in our department's conference room, and as always, she began the conversation by asking about my wife and kids as well as my family back home in Illinois. Wanda is smart and talented, but she is also a sincerely nice person. She said that Steven was thinking about pursuing a career in the fire service, and she wanted my advice on education, his best next steps, and where his talents, as I saw them, might fit related to a fire service job.

Of all the tasks I get to do as a fire chief/CEO, one of my absolute favorite is helping someone figure out how to pursue their passion through their work. I love to help them see a vision of what is possible with hard work and a desire to do something important. I like to help them order their plans and to then celebrate with them as they accomplish each small achievement as they climb their own individualized ladder of success. I'm never too busy to help in this area, and Wanda was asking me to do what I love. She was asking me to think about career planning and succession planning.

I was excited to hear about Steven's desire to pursue a full-time fire service career. I remember giving Wanda an overview of the global U.S. fire service and the differences of working for agencies in different parts of the country. We explored the variations in types of departments and the specifics of combination, fully career, and metro departments. I stressed the importance of finding a department that aligns with Steven's personality and abilities. I also remember urging that he pursue a degree in fire administration and that he should absolutely plan on attending paramedic school. But what stands out most in my memory is me saying that I could see him down the road serving as my assistant fire chief and helping me run King Fire Department. I even remember saying that he might be the one to replace me when I retire. I am struck now by the prophetic nature of this conversation.

We ended the meeting, and as she left, I remember hoping that my advice was sound and that it would prove helpful. I committed in my mind to try and give Steven opportunities at King for training as well as experiences where he could develop his skills as both a firefighter and fire service administrator. The truth of the matter is, Steven is naturally gifted, and he was going to bloom wherever he was planted. Although I would have been happy to see him succeed in any fire service position within any agency, I was really hoping to see him grow deep roots at King Fire Department.

Steven attended school and worked in various positions as a volunteer, part-time, and career firefighter both within King Fire Department and at some of the nearby neighboring emergency service agencies. He received degrees in fire service administration, emergency preparedness, and fire protection technology. He went to paramedic school and worked for the county ambulance service and did a stint with two different county fire marshal's offices as an assistant fire marshal. While in these roles he even served as fire chief of the Walnut Cove (NC) Volunteer Fire Department before being appointed as King's fire chief (fig. 5–1).

I left King in 2002 to return to Illinois and to become fire chief at Hanover Park. This move had not been expected on my part, and I really thought that I

Figure 5–1. Chief Steven Roberson (left) providing direction while working the command post (photo courtesy of City of King Fire Department).

would get to personally spend more time with Steven. But the fact of the matter is that he is tremendously driven and talented, and he worked very hard to excel in his pursuits. When I left, he was not yet ready to be fire chief. The city filled the position from outside the organization, but seven years later he was given the opportunity to step into the top spot and lead his hometown department.

I often think about my conversation with Wanda and my comment that he might be the one to replace me as fire chief. Although the city went outside initially due to the limited experience of our newly formed career department, I am incredibly proud that they stayed the course, worked to develop one of their own, and were ultimately able to fill the top spot with someone from inside. I think this is a great example of succession planning.

SUCCESSION PLANNING AS A HOT TOPIC

Succession planning is a frequent topic of conversation in today's fire service. Whether I am speaking at a conference, providing training at a local department, or simply visiting with a group of fire service personnel, succession planning always seems to make its way into the conversation. Sometimes these conversations are with department leaders who are asking questions about how to do succession planning. In most of these scenarios the leader is nearing retirement and is beginning to think seriously about who will fill the role of fire chief/CEO once they exit the organization. Occasionally, I am asked about the best way to develop staff to fill senior leadership roles and what needs to be done to get them interested in taking promotional exams. Sometimes I hear stories of leaders who are working to engage their personnel in succession planning activities—and no interest seems to exist.

Other discussions are with firefighters and company officers complaining that their leadership is not doing enough related to succession planning. They complain that "Da Chief" does not involve anyone else in running the organization and does not share information. They then commonly launch into a story about how the chief made a poor decision, seemingly in a vacuum, that has turned out to be negative in the eyes of the staff. "If he would have just allowed others to be involved or sought some input, things would have worked out okay."

At other times these discussions with firefighters and company officers revolve around how no one is interested in moving up and taking on senior leadership roles. This is usually followed by an explanation of how their members are good firefighters and do good work on the street but how they want nothing to do with the politics of running the department. In the end though, they usually get back to their perception that if the chief would just do something different, all would be right in the world of the Hooterville Fire Department. But when asked what the chief needs to do differently . . . well . . . they don't exactly know.

As I ponder the various aspects of these discussions, both the ones from the fire chief/CEO as well as those from the firefighters and company officers, I have come to believe that

- the fire service does not really understand the term succession planning,
- junior staff believe that leadership should be doing "something" regarding succession, but
- neither side really understands their specific roles nor knows what successful succession planning should look like.

THE PROBLEM

Succession by its nature focuses on a line of progression designed to pass the torch of responsibility to those next in line. The fire service works hard at developing the abilities of our underlings to receive the torch and learn to manage emergency incidents, but we frequently fall short in developing their organizational leadership skills so they can lead the department into the future (Haigh, 2010).

Generally, when we hear the phrase "succession planning," our mind immediately jumps to senior organizational leadership skills and the need to equip up-and-coming personnel with the tools required to manage the organization as part of command staff. But equally important is development of the knowledge base and skill set needed to manage emergency incidents, whether at the company level or the front office. Succession planning needs to occur at all levels of the organization, and it is the responsibility of every member to be doing succession planning with those in their immediate area of influence. My suspicion is that most departments are doing a level of succession planning and probably don't recognize that the work that is occurring is just that. I think they attach the skills of senior organizational leadership to the phrase "succession planning," and although it applies to this area, which is generally where we fall short, the reality is that succession planning is much more than developing the next fire chief/CEO.

So why do we fall short related to teaching senior organizational leadership skills?

I think that both sides—fire administration (i.e., fire chief/CEO, senior command staff) and labor (i.e., firefighters and company officers)—realize that developing the nonoperational skills of future leaders is important, but neither seem to know exactly what to do or how. They throw around lots of blame, but in the end, both sides are really at a loss in this area of succession planning.

I liken this to knowing that something is wrong with your car but not being able to figure out what to do to fix it. You know enough about cars to know how to fuel it, check the oil and tire pressure, and add washer fluid, but everything else about how a modern vehicle works is a general mystery. When it starts making a funny sound or one of the idiot lights illuminates on the instrument panel, you feel kind of helpless. You are at a loss and need to find a solution. You need a mechanic.

Likewise, regardless of your position within the organization, you know that a successful department needs finances, personnel, and equipment (e.g., fuel, air in the tires, and washer fluid), but you are stymied by the complexities of making it all happen within the context of a governmental institution, coupled with the myriad of rules generated by the alphabet soup entities (OSHA, DOL, DOJ, NFPA, etc.).

When your car breaks down, it is simple. You need a mechanic to figure it out and fix it. When it comes to succession planning related to organizational leadership skills, you need someone who both understands the complexities of fire administration (i.e., CEO skills) and can then teach these skills to adult learners. You need a mechanic in the front office who can take apart the complexities of administration and provide teaching that encompasses classroom training, experiential learning, and mentoring. All things that take considerable time and attention.

I think it is important to remember that we do all these things well for operational tasks. We have plenty of operational mechanics who can break down and teach the steps associated with forcing a door, stretching a hose load, pumping an engine, venting a building, and searching a structure for victims. We just need the same type of mechanic for the myriad of administrative skills involved in running a department.

One problem is that we don't have trained mechanics within our organizations to provide this specialized education at the upper levels. Compounding this problem is that most of our fire service training institutions do little to teach CEO skills to those who aspire to learn them. We don't talk about the needed skills, don't teach classes that focus on them, and in many ways downplay their importance. Yet the higher you climb the organizational structure of a fire department, the more critical it becomes to focus on these skills. Through our failure to acknowledge and address this issue we often leave our organizations in a bad spot related to upper management succession.

WHY WE STRUGGLE WITH SENIOR LEADERSHIP SUCCESSION PLANNING

I think the challenges associated with succession planning are multifaceted and complex. Here are a few of the issues as I see them:

- People are attracted to this profession because of the excitement of being an emergency services provider. Most didn't join the fire service hoping to drive a desk. They would rather run calls and do the exciting aspects of our business. They don't see the administrative tasks as stimulating and therefore have little passion to learn these skills.
- We typically don't recruit and hire firefighters who have an equal interest in both riding a rig and doing administrative and organizational leadership work. When we recruit, we generally focus only on the street-level work and completely disregard any focus on the administrative aspects of the job. Based on this, it should come as no surprise

that we struggle in this area as we try to develop interests and skills in our current staff, which were skills we never assessed as part of the recruitment process. Bottom line, it is not their area of giftedness, and they don't have a passion to do the work.
- We often ostracize members who are hired who have an interest in the administrative aspect of the job. We talk regularly about the brotherhood and sisterhood of the fire service, but let someone get appointed who thinks differently from us or has a skill base beyond street-level functions and we quickly label this firefighter as a pencil pusher and see no value in having them on the team. We can be incredibly cruel to these individuals instead of embracing the fact that, if molded in the right fashion, they are the ones who possess the skills to effectively lead the organization in the areas where the typical firefighter does not have the skills or interest. Sometimes we even push them out of the department simply because they don't fit the typical firefighter model.
- Many of our current staff and officers see succession planning as a threat. Existing employees chastise and work to discredit those who are driven to learn and build a career beyond their current position. Sometimes consciously, but more often subconsciously I believe, we beat down our bright and shining stars, discouraging them, holding them back, and throwing up roadblocks to prevent them from moving ahead. The prevailing but unspoken thought is that if they begin to have an impact on how the organization operates, beyond pulling hose and swirling toilets, they somehow are making others look bad. Those who have no interest in advancement or who simply lack the ability to progress work hard to diminish these administratively talented members simply out of fear or jealously (Haigh, 2010).

Because they were never formally trained in the administrative and organizational leadership skills needed to manage an organization, many chief officers are not comfortable with their own abilities to lead the organization and are therefore not equipped to develop and teach the skills needed to those coming up behind them. Similarly, many feel that teaching someone else what they know will somehow diminish their authority and make them less important. Weak leaders often try to play things close to the vest and rule with an iron fist. This builds their status based on fear and making sure they know more than others. Generally, this comes from their own fears and insecurities about the job they are doing and about limitations within their personal abilities (Haigh, 2010). I like the words of the late Chief Alan Brunacini in his legendary book *Fire Command:* "Be careful of people who attach status to knowing things you don't" (Brunacini, 1985).

- The jobs that involve administration and organizational leadership are not attractive in today's fire service culture:
  - For career firefighters, they are not enticed by the schedule change that forces them into a 40-hour week from their typical 24-hour shifts. Sometimes this is due to their "side job," which would be significantly impacted by a five-day-per-week schedule. Others use their nonshift days to provide care for their children while their spouse works full time.
  - For volunteers, they recognize that in order to effectively run the organization, it is a full-time job. This full-time job is on top of their "real job," and there is limited financial incentive (or other benefits) to do the work. Also, I continue to be amazed at how badly some volunteer chiefs are treated by their community's leadership. They are giving freely of their time and talents but are treated by managers and elected officials as if they are the enemy.
  - Many personnel who are capable of serving in the top spots see the workload, political pressure, stress, and toll it takes on the leader and their family and are simply not interested in accepting that weight.
  - They see that leaders live in the proverbial world of "between a rock and a hard place" as they try to navigate between the will of the elected officials and community/district leadership and that of the membership/union.
  - The money and benefit differences are not enough to make the positions enticing. Most firefighters today can work a few overtime shifts and make the same or more money than the officers in senior leadership roles. This overtime issue can become problematic when trying to entice quality and capable members to step into an FLSA-exempt position. In many cases by accepting the promotion they end up taking a significant reduction in overall pay.

- Volunteer and paid-on-call agencies are struggling with recruitment and getting members to join. Once you get them, due to the time demands of jobs, family, and the needs of the department, keeping these critical assets is very difficult. Many times, these members join the department before they have a family, but as they move into the next phase of their life and family demands increase, they find

that they no longer have the time to volunteer. I have also seen more than one department with a rockstar volunteer who has the marketplace skills that can easily be converted into what is needed to successfully lead the department, only to have them snatched away by their real job as they are transferred to a different part of the country or offered a better position that prevents them from continuing to volunteer. Likewise, few volunteer departments today have not experienced the loss of a member who gets hired as a career firefighter and who has the skills that could be built into a future senior command staff position within the volunteer ranks, but either due to the rules of the career department or the rules of the union, the member is prevented from continuing to serve in their hometown volunteer department.
- Sometimes I see unions representing career firefighters block succession planning activities. Because they don't fully understand succession planning, they believe that leadership is trying to select individuals for promotion and thereby circumvent the established processes that give all members fair opportunities for future positions. In reality, when succession planning is done correctly, the process works to develop candidates so that they have the best opportunity for success both in the testing as well as in job performance once promoted. It is not a process of selection but rather a process of learning.

ORGANIZATIONAL VALUE

For succession planning to be effective, it needs to be viewed as a global organizational value. It begins with recruiting and hiring the right employees and creating opportunities that allow them to grow and develop while simultaneously providing strong mentoring and encouragement. It is not about who will be advancing in the next three to five years, but rather it includes a much broader view of employees and how they individually fit into the organization's current and future needs. For continued department success and stability, succession planning is one of the most important aspects of leading an organization. As leaders, we need to see each employee as an individual and then learn how to assess their unique gifts and talents so we can use these for the benefit of the organization.

I think that a great many leaders start to seriously think about succession planning as they begin to contemplate their exit from the organization. Unfortunately, if this is the model they are employing, I would argue that they are far behind in preparing their personnel for sustained organizational success. Succession planning is not about any one leader/fire chief/CEO but rather about all positions and all personnel who are members of the department. Personnel at all ranks are going to come and go throughout the life of an organization. Some are going to get reassigned, promoted, retire, or leave the organization for another opportunity. One absolute fact is that all organizations are in a constant state of flux related to personnel. Likewise, all personnel are at different stages of development and at different points in their career. Succession planning needs to be applied to all positions and to all personnel in order to meet them where they are.

EVERYONE'S RESPONSIBILITY

An example of how succession planning is supposed to work.... The firefighters and officers of a fire company, let's call them Engine 16, need to continually be thinking about the members, their stage of development, and what skills they need to mature in order to be stronger members of the team. The senior members need to pay particular attention to the junior members, working to train them on the various tasks performed by the company so that when they go out the door, they are prepared to perform with excellence. They need to instill company pride to be eager to wear T-shirts with Engine 16 printed in bold letters on the back and to work with a team that shows others they are great at their job. They need to become the team that they want responding to their family's emergencies. This is achieved by senior personnel who are committed to developing the more junior members. It is not the responsibility of the training division; it is the company's responsibility to own this aspect of employee development. The structure of senior members training and mentoring the more junior members is succession planning in its simplest form.

Likewise, the company officer of Engine 16 needs to be working to train and mentor the senior members. This will allow an experienced team member to be prepared to step up and cover for the officer when they are absent. The officer should be delegating a variety of duties to this member, including some of the administrative work associated with running a company and firehouse. This is not "passing off work" but rather giving the member the opportunity to learn under the tutelage of the company officer. If the senior member is preparing for promotion, it is the company officer's responsibility to help this member develop the skills needed for this next step in their fire service career. This is also succession planning.

Besides training and mentorship, the members of Engine 16 should also be thinking about and assessing other department personnel who might make excellent future members of the company. If I were an officer on Engine 16,

I would always be on the lookout for talent that I could somehow grab and make part of my company.

This succession planning model should extend from the company officer to the battalion chiefs/district chiefs/shift commanders as well. Their job is to develop the company officers under their command and to prepare those interested in future promotion. This work normally involves teaching both the tactical aspects of managing an emergency incident as well as the management and leadership skills needed in a command level role. Just like with a company, the battalion chiefs/district chiefs/shift commanders need to work to build their team into one that attracts the best firefighters and officers.

Organizationally this focus on mentoring and development needs to continue from one rank to the next up through the position of fire chief/CEO. It needs to be an intentional part of how the organization operates and something that is done daily with each interaction, meeting, and training. Some of the best succession planning occurs around the kitchen table or while sitting on the park bench outside the bay doors. Sharing of knowledge and mentoring needs to be an organizational cultural value. Storytelling is a great way to accomplish this. If you want to be great at succession planning, tell stories of how you learned (including your failures). The stories will stick and will become the basis for learning by others. Withholding information and failing to develop members coming up behind you will create a future void of knowledge that will cause the organization to falter. Succession planning is not a special class or training program; it is the way your department develops staff day in and day out.

MAKING SUCCESSION PLANNING A CULTURAL VALUE

I came to Hanover Park in July of 2002 as the outside fire chief. The department had recently become part of the village after being a separate fire protection district. The merger had occurred due to financial challenges that had been compounded by years of poor leadership. It's not that the department had bad personnel—in fact just the opposite. It's just that no one had ever been developed to lead the organization from the position of fire chief. This lack of development caused the leadership to make poor decisions, which ultimately led to the demise of the district. My joining Hanover Park was the second time that I have been hired as the outside chief (King, NC, in 1995 and Hanover Park, IL, in 2002).

I realize that fire chiefs are routinely hired from the outside and that my experience in moving between two different agencies to fill the top spot is not necessarily unique. The experience, however, of being the "outside guy" within two different agencies provided me with insight as to why departments make this decision along with the good, bad, and ugly that goes along with this type of change.

In general, I feel that hiring an outside chief carries substantial risk. Departments, like families, have their own unique values and culture. Sometimes this culture is good and well suited for the community it serves. Sometimes the culture needs to be tweaked, but rarely does it need a wholesale change. When bringing in a new CEO who has no history with the organization (i.e., limited knowledge of the culture) and often some baggage of their own, the attempt at blending the outside (i.e., the new chief) with the existing department can be difficult.

I try to coach new outside chiefs first to not break anything until they have a good idea as to what is really going on. It is important to realize that the fire chief/CEO job is not that of the U.S. presidency and the media is not normally tracking and reporting on your "first 100 days." Therefore—move slowly! Unless some unique circumstance or situation exists, I think it best to watch, listen, and work to figure out the lay of the land as well as the players involved before making any major changes. Moving too fast can cause a disaster, and once something good is broken, it is hard to put it back together.

Secondly, it is imperative that you become a member of the new organization and not try to make it into the department you just left. The members of your new department will have some interest in learning some facts about your old place, but a little bit of sharing goes a long way.

Prior to my acceptance of the job offer, I had numerous interactions with village leadership, department members, and the executive boards of the International Association of Fire Fighters (IAFF) Local 3452 and the Service Employees International Union (SEIU) Local 73. I made it clear that if I accepted the position, I would evaluate the department and make changes where necessary, but I would spend the majority of my time working with them to create a cultural value where the department implements succession planning in all aspects from recruitment of new members to promotions. My long-term intent with this focus would be to position the organization so that future fire chiefs/CEOs could come from inside.

FARM TEAM

Once I accepted the job offer and was sworn in as the fourth career fire chief in the history of the department, my first step in the succession planning process was to begin strengthening our part-time firefighter program. Hanover Park utilizes a staff of part-time firefighter/EMTs to fill full-time shift vacancies occurring from annual leave, Kelly Days, sick leave, and scheduled training furloughs. All of these members are certified firefighters, and the majority are licensed paramedics. Besides providing full-time backfill, these members are also responsible to staff the Power Shift, which is a peak time company that works Monday through Friday. This company is used to conduct most of the annual business license and safety inspections as well as the ISO-required annual prefire plans. Power Shift personnel also cross-staff an ambulance during periods of heavy call volume.

More than 80% of the department's current full-time personnel started their career with Hanover Park as a part-time firefighter. This percentage has been fairly constant, and at different times in the department's history, even higher. I realized quickly that this part-time group should be viewed as our "farm team" and that they could, if managed correctly, have a significant role in the long-term success of the department's succession planning.

In his book *Developing the Leaders Around You*, John Maxwell wrote:

> Spend more effort on the "Farm Team" than on "Free Agents." In major league baseball, teams generally recruit players in one of two ways. They either bring players up from their own minor league farm teams or go outside the organization in search of free agents. Time after time, baseball fans have seen their teams bring in expensive free agents with the expectation of winning a World Series. Time after time they are disappointed. The Farm Team model brings in the best underdeveloped players. (Maxwell, 1995, p. 28)

Maxwell recommends that leaders focus on coaching and developing farm team members into the players needed to win the World Series. He suggests doing this by the following steps:

1. Invest time and money in these members.
2. Commit to promoting from within whenever possible.
3. Show people that personal and professional growth within the organization is not only possible but actual.

Although this farm team analogy could be applicable to all members as they progress through the various ranks, I thought it best to begin succession planning at the recruit level. Building from the ground up gave me a firm foundation on which to begin constructing a cultural change that focused on succession. For Hanover Park, this is the part-time firefighters.

The department's part-time members are unionized and are represented by SEIU. When I arrived, this group had just recently organized in response to the recent merger of the department into the village. They feared working for the village, feared what the outside fire chief might do, and were scared for their continued existence within the department. With the help of IAFF Local 3452, the part-time group formally organized and was ready to begin negotiating its first collective bargaining agreement. Their goal was simple: ensure their continued existence.

My goal mirrored theirs, but I also wanted to expand this goal as it related to succession planning. I wanted to make our department and our program an attractive option for candidates looking to begin their fire service career. I wanted Hanover Park to become the first-choice organization for these yet undeveloped players. To do this we bolstered their pay, provided language on how they were to be trained, agreed to pay for uniforms and protective gear that met relevant National Fire Protection Association (NFPA) standards, and agreed on a disciplinary process similar to what the full-time unionized members had in their IAFF contract. We also agreed to add hiring policies that gave them preference points when competing for vacant full-time positions. Negotiations went relatively well, and we quickly solidified a new contract.

With the part-time union contract now in place, along with the new rules for full-time hiring preference points, we began to focus significant attention on part-time firefighter recruitment. We decided to apply the exact same testing and hiring standards used for full-time personnel to our part-time process. This would ensure the quality of candidates and would eliminate any concerns about giving these members preference points in full-time hiring. Our goal was to recruit and hire the best underdeveloped players and build them into a winning team.

DUAL ROLE BATTALION CHIEFS

The next major succession planning change came as we changed the long-time position of lieutenant/shift commander to the rank of battalion chief. This move officially gave the department members another position to strive for while solidifying a command team within fire administration. Because we had talented members, even though they lacked some training and skill development in organizational leadership, we were able to promote from within to fill this new rank. These newly appointed chief officers were not only grateful for the opportunity to lead in this new capacity, but they were each committed to working hard, learning, and doing all possible to grow into their new role and to meet the expectations of the new fire chief.

Because we are a relatively small department, each of these new battalion chiefs was assigned to not only command their respective shifts but also to also lead a major division of the department (operations, training, and EMS). Within each of these divisions, coordinators were assigned to further subdivide the workload. Coordinators are personnel who hold a rank less than that of battalion chief and in many instances are firefighter/paramedics. They were assigned based on an area of interest, special expertise, and willingness to do the work. In many cases, the best coordinators were not officers (i.e., lieutenants) or at least not yet. Their job was help the battalion chiefs run the divisions. From the succession planning standpoint, these coordinators were now engaged in learning how to administratively manage an assigned area of the department.

A unique aspect of this division of work and responsibility is that all battalion chiefs and their coordinators are actually 24-hour personnel who complete their administrative duties in addition to their normal shift work. They fit their administrative responsibilities into their regular work shift and are given latitude, when necessary, to work overtime on their days off.

A word on this overtime: when this plan was first devised, village leadership was concerned that the battalion chiefs and the coordinators would exploit this available overtime. Although this is a valid concern, I didn't think it would happen. My sense was that the newly promoted battalion chiefs, along with their coordinators, were excited to be given this new opportunity to lead. This had been missing within the organization, and this change brought a renewed sense of passion and drive. They knew that village leadership was concerned about this issue, and they also knew that I had stuck my neck out to get the promotions approved. I was now trusting them with my personal well-being as the new fire chief, and this trust set a tone of self-policing of their spending in order to protect me. It was a calculated risk, but leadership is sometimes risky.

To put numbers with this situation, a division chief here in the Chicagoland fire service market, with salary, benefits, and pension costs, would be about $180,000 annually. Without the coordinators and battalion chiefs doing dual duty, Hanover Park would need to have three 40-hour administrative personnel to manage the three divisions (operations, EMS, and training). The annual cost for this level of staffing would be more than half a million dollars annually ($540,000). Using our model, with the applied diligence of the battalion chiefs and coordinators, we spend around $75,000 annually in overtime to make our program operate. I would argue that getting the same level of leadership without spending the additional $465,000 annually is a pretty good bargain for the taxpayers of our community.

As the years have passed, this program has fallen into the groove of "this is how we operate." The original group of battalion chiefs retired, as well as the next group who followed them. We are now into our third generation of battalion chiefs. In fact, none of the current battalion chiefs were even part of the department when the original positions were created and the dual duty roles envisioned.

The current group of battalion chiefs all came up through the department and served as coordinators, got promoted to lieutenant, took on additional administrative duties, and then ultimately became battalion chiefs (Hanover Park does not have the rank of captain). What has been interesting to watch is how newly promoted lieutenants, after serving as coordinators, seem to easily and quickly transition into their new role. It is almost as if they had been doing the job long before they got promoted. They generally make the transition almost seamlessly. This has me thinking that we are on the right track related to succession planning!

SHARING WHAT WE LEARNED

Our personnel began to share with their buddies in other agencies what we had going on, good or bad (remember—telegraph, telephone, tell a firefighter). This led to me receiving questions from my peers as to what exactly we were doing. Also, because of the robust auto aid system here in the Chicagoland area, my peers could see how our personnel functioned at emergency incidents, but they were curious about what was happening administratively. To explain my succession planning strategy, I developed the following chart to explain that we were basically asking our personnel to think and perform administratively one step above their current rank and pay grade (fig. 5–2).

This practice, partnered with sound mentorship, seemed to work, and we were making significant progress moving the department in the right direction for functional excellence.

PREPARING THEM FOR THE JOB: TRAINING

In order to prepare personnel for administrative roles, both today and into the future, you need to train them. But what does a training program look like? This is a question we continually wrestle with. The fire service generally wants a program that is all neatly packaged and tied up in a bow. I get calls all the time from fellow fire chiefs asking for my "succession planning program." They seem shocked when I tell them that I don't have a "program" but rather a bunch of different stuff that when linked together helps to prepare personnel for future leadership roles.

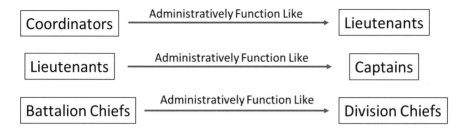

Figure 5–2. This strategy shows personnel how to think and act one step above their pay grade.

I think this "program" idea comes from our fire service certification standards and the instruction that you need to meet certain objectives in order to get the certificate. Unfortunately, administrative and leadership skill development and learning is much more complicated than just meeting a specific objective. What works in one situation may not necessarily work in another. Therefore, you need to understand the theory behind the issue, the people involved, the overarching desired goal, and about a hundred other issues that all seem to converge as you try and make the best decision on how to do something. Additionally, as time passes, the work tends to change based on newly available information and changing cultural trends. When training in this arena you don't need a program, but rather you need to teach members how to think.

As leaders working to provide succession planning, I have come to believe that we need to be in a constant state of assessment. I like to ask the following questions as we work on determining how best to train people:

1. What skill sets will be required for future department leaders?
2. What skills, knowledge, and experiences are required for each level within the department?
3. What steps can the department take to prepare future leaders?

None of these questions seem all that difficult until you begin to break them down to determine action steps.

Question 1, as an example, focuses on skills of future leaders. Stephen A. Miles and Nathan Bennett, in their *Forbes* magazine article "Best Practices in Succession Planning," wrote, "Only in the rarest cases will future challenges require the same skills that worked in the past" (Miles & Bennett, 2007). None of us have a crystal ball that we can use to predict the future, but we certainly should be asking ourselves what the department will need to look like and do

20 years down the road. Although the picture will not be entirely clear, we can make predictions based on emerging trends (Haigh, 2010).

As an example, when I was coming up as a future leader of the fire service, I was advised by my mentors to pursue degrees in fire science and public administration. I think that this made sense 30 plus years ago, but based on what my job today actually entails, I would give a future fire chief different career advice. The formal education that would be most beneficial to me in today's environment would be a law degree. If I had to do it again, I would have gone to law school instead of getting my master's degree. Not that I want to practice as an attorney per se, but so much of my daily job focuses on employment laws, contracts, and statutory regulations. Organizational liability is a huge issue in our current litigious environment. I used to spend so much time meeting with and discussing issues with our various attorneys, that it would be hugely beneficial if I was one myself.

As I ponder my advice for the future based on emerging trends that I see in our business, I think formalized education and degrees in health care administration, public policy, technology (data analytics, apps, game design, virtualization), and human relations might make the list for a future fire service CEO. I am sure there are also numerous others. The key here is to make sure that as we mentor our people we try to give them the best advice possible and help them think globally about where the fire service will be as we head into the future. What is relevant today may not be so relevant in the future.

Question 2 looks at the work done within each respective level of the organization and what someone would need to know and be able to do if they were to fill that role. What experiential learning needs to occur for them to be successful in each position?

At Hanover Park we worked hard over the years to provide members with training opportunities designed to broaden their knowledge base beyond just street-level emergency response skills. A few examples include hosting trainings with national speakers and experts in various management topics. We sent members to the National Fire Academy, conducted department-specific workshops on policies and procedures, and encouraged members to further their formalized education through tuition reimbursement programs. We have also tried to give them opportunities to use this new knowledge base through work as a coordinator.

Question 3 looks at what an organization can do to help prepare future leaders. A key for this area revolves around mentoring and coaching. Mentors and coaches should begin teaching members new skills and tasks and, as they succeed with the small things, reward them with more and more responsibility. Coach them through the process and allow them to make mistakes, but do not set them up to fail or give them more responsibility than they are ready to handle. The goal is to teach and allow them to grow (Haigh, 2010).

CAREER DEVELOPMENT PLANNING

A critical component of succession planning is talking with your personnel to determine their career goals. Once you have an idea what they are thinking and where they want to head, you can then begin to help them create a learning and development plan that focuses on these goals. After years of working with firefighters, I find that they often have an idea of where they want to go but generally have no clear road map for making the trip. This is where mentors play a critical role in helping to draft a plan that is both attainable and a stretch. Stretching is important. Goals that are easy don't do much to help the member grow and develop. We all need to periodically be stretched in order to achieve our full potential.

When meeting with members to discuss career planning, I like to use the National Fire Academy's Personal Analysis and Development Plan. This plan, used as part of the Executive Fire Officer Program, asks questions designed to facilitate a path of self-reflection in several categories. Self-reflection is critical in order to decide what is most important to the member and what they are willing to sacrifice in order to achieve their goals (achievement of goals is always associated with sacrifice). The document walks the user through five specific categories of thinking:

1. Self-reflection
    - Where am I now?
    - Who do I think I am?
    - What opportunities, challenges, and potential contribution do I see in my future?

2. Values and goals
    - What is most important to me in my life?
    - What are the most important goals in each of my life areas?
    - Family life?
    - Work life?
    - Social life?
    - Self?

3. Learning plan
   - Where do I need to grow?
   - How can I make the best use of my time in order to meet my life goals and my learning goals?

4. Feedback (What do my mentors see in me?)
   - Are my learning goals and learning plans realistic?
   - Have I identified all opportunities and resources available to me?

5. How specifically am I going to commit to a learning plan?

Once they have worked through this self-reflection, we then begin to translate this data into a written and measurable learning plan. The plan should address the following:

1. What are your desires and passions?
2. What are your core job strengths?
3. What career options exist based on your skills?
4. Set goals and realistic timelines for career objectives.
5. Implement your plan (i.e., go to work on getting it done).
6. Accountability:
   - Meet with a mentor on a regular basis to review progress.
   - With the help of a mentor, adjust and modify the plan based on what you are learning (learning about self, new skills, new knowledge).

7. Plan for next career step: after meeting one goal, start the process for the next.

UNDERSTANDING YOUR ORGANIZATION'S SITUATION

Besides the focus on developing individual staff members, department leadership needs to have a global picture of the organization. This allows leadership to see any potential pitfalls that might not be readily apparent. I utilize a spreadsheet to track each individual member of the department and their succession planning strategy based on conversations related to their career development planning. On my spreadsheet I list out the various positions I am tracking (e.g., command staff, company officers, firefighters) along with the name of the individual currently filling the role. Data is then entered as follows:

- Hire date
- Date of promotion to current position
- Date of entry into the pension system (due to the part-time program, their hire date and pension date may be different)
- Pensionable years of service
- Age
- Retirement status
    - Based on conversations with the employee, are they planning to retire within one, three, or five years?
    - If they anticipate it being beyond five years, no retirement status is listed.
    - This status report does not tie the employee to retirement; it merely provides helpful information in order to accurately assess the organizational succession planning picture.
- If the position is slated to be filled via a tested promotional list, how many members are on the current eligibility list?
- Criticality: Will a newly promoted member need to "hit the ground running" or be fully functional within six months?
- Based on current membership, how many members are ready now to fill a vacancy within a given position? (This is a judgment call based on my assessment of the team members.)
- How many staff will be ready to fill a vacancy within that position in the next one to two years? (This is a judgment call as well.)
- Based on information shared during career development planning, each employee is given a "succession planning priority" based on their desire to be promoted, their individual capabilities, and their likelihood in scoring high enough on a promotional exam to be promoted (fig. 5–3). (This gives me an idea on where to spend my limited training budget in order to generate the greatest gain for the department.)

None of this information should be a secret. My desire is to have members weigh in and help identify their status in each of the areas. I have found that this helps to generate ownership and assists them in seeing the bigger picture of how they fit overall into the long-term success of the organization.

Hanover Park Fire Department Succession Planning Worksheet												
Position Title	Incumbent Name	Hire Date	Promotion Date to Current Rank	Entry Date into Pension System	Pensionable Years of Service	Age	Retirement Status	Number of Staff on Promotional List	Criticality	Number of Staff Ready Now	Number of Staff Ready in 1-2 Years	Employee Succession Planning Priority
Fire Chief												
Assistant Chief												
B/C EMS												
B/C Operations												
B/C Training												
Chief of Inspectional Services												
Lieutenant												
Lieutenant												
Lieutenant												
Lieutenant												
Lieutenant												
Lieutenant												

Retirement Status:	Succession Planning Priority		Criticality:
A = Retirement likely within 1 Year	High	1	1 = Must "hit the ground running"
B = Retirement likely within 3 Years	Moderate	2	2 = Must be functional within 6 months
C = Retirement likely within 5 Years	Low	3	
	Will not promote further	0	

Figure 5–3. Succession planning worksheet (courtesy of Hanover Park Fire Department).

THE REALITY OF PEOPLE

I have seen it repeatedly . . . the high performing fire officer who is hugely gifted and has tremendous promise, the one who just seems to instinctively know how to do the job. They cast a vision that their team can follow and that their people can trust. Others try to learn from them, imitate them, and hold them up as an example of how leadership should be done. Their abilities seem to come naturally, and people are instinctively drawn to them. As you watch them function, it is obvious that they would do a tremendous job in the top command level positions of any organization, not just in the fire service, but for whatever reason, they are unwilling to make a grab for the proverbial brass ring.

In these cases, the decision to stop climbing the organizational ladder is usually deeply personal. Although they have the ability and would be the individual that most folks would want elevated into the top organizational spots, it is just not something they are willing to do. Some of these members have caused me great frustration over the years. I see their abilities, and I often find that I want more for them than they want for themselves. I see the great promise they have and how their skills would greatly impact the global fire service. They just won't take the chance.

Likewise, I see talented firefighters who stumble their way through their careers hoping for a promotion and who halfheartedly try to get the job. If they would simply make a sincere attempt to do more than the minimum, their sheer talent would likely make them outstanding. Yet this doesn't happen, and they never realize their full potential because they don't push themselves beyond their comfort zone. I generally think of these personnel as lazy (although this may be too harsh a term), unfocused, and simply unwilling to work very hard. They lack the inner drive to succeed mightily.

Then there are those who lack the natural talent, are not the most gifted, and were certainly not the most popular and not the straight A students in high school. They are the ones who have had to work harder than everyone else to achieve the same results, the ones who, when others quit for the day, still have their head stuck in a book, or the ones quietly sitting on the tailboard of a rig practicing knots into the late hours of the evening. These are the ones who have a personal vision for what they want to accomplish, the ones who hold themselves to the highest standards, and the ones who, when given an opportunity, will give an all-out effort, every single time.

As we work on organizational succession planning, it is imperative that we remember that we are dealing with people. Some of our people are hugely talented, some lack drive and motivation, and some have more passion and heart than most Olympic athletes. Although we as leadership can clearly see how certain individuals might fit well into the various blocks on our organizational succession planning chart, the fact of the matter remains that they themselves need to want to fill those roles.

It is only natural for an invested leader/fire chief/CEO to feel like the rug has been pulled out from underneath you when you find that all of your work related to planning, recruitment, hiring, and employee development designed to put the right people in place at the right time to ensure seamless succession is not working because the people involved don't share the same vision for themselves that you do. You must remember that they are people and they have the right to set their own personal work and life plans—and that is okay. All we can truly be responsible for is ourselves, and we need to realize that in the end, although our goal has been to promote from within the organization, sometimes our own team members, based on their personal life desires, will force us to look outside for players to fill some of our key roles. This does not mean that you have failed in your succession planning work.

I can't imagine a case when a leader focused time and attention on succession planning where recruits were not developed into firefighters, where firefighters were not developed into senior firefighters, where senior firefighters were not developed into company officers, and where company officers were not developed into command officers. Succession planning is more than working to fill any one particular position. It is about the entire team, and if the majority of your team is growing and developing, then you should feel encouraged that you are getting your succession planning activities more right than wrong.

REFERENCES

Brunacini, A. V. (1985). *Fire Command,* 1st ed. Quincy, MA: NFPA.

Haigh, C. A. (2010, August). Legacy of Leadership: Hiring the Right People and Developing Them for the Future. *Firehouse*.

Maxwell, J. C. (1995). *Developing the Leaders Around You*. Nashville, TN: Thomas Nelson.

Miles, S. A., & Bennett, N. (2007, November 11). Best Practices in Succession Planning. *Forbes*. https://www.forbes.com/2007/11/07/succession-ceos-governance-lead-cx_sm_1107planning.html#41e95d66668d.

QUESTIONS FOR FURTHER RESEARCH, THOUGHT, AND DISCUSSION

1. Where within your organization is succession planning occurring?

   a. _____
   b. _____
   c. _____
   d. _____
   e. _____

2. What are some of the stumbling blocks your organization is facing related to succession planning?

   a. _____
   b. _____
   c. _____
   d. _____
   e. _____

3. What could be considered your organization's farm team (future employees)? If you do not have one, can you build one? Provide a general overview of what this might look like.

   _____
   _____
   _____
   _____
   _____

4. In what areas is your organization giving members at lower ranks the opportunity to be involved in leadership roles? How can this be expanded or strengthened?

   _____
   _____
   _____
   _____
   _____

5. Organizational assessment questions:

    a. What skill sets will be required for future fire chiefs/CEOs?

    _____
    _____
    _____
    _____
    _____

    b. What skills, knowledge, and experiences are currently required for each rank/level within your organization? How do you anticipate this to change in the coming year?

    _____
    _____
    _____
    _____
    _____

    c. What steps can your organization take to prepare its future leaders?

    _____
    _____
    _____
    _____
    _____

6. Think critically about your own career development plans:

    a. What opportunities, challenges, and potential contributions do you see in your future?

    _____
    _____
    _____

    b. What would you identify as the most important things in your life? How do they correspond with your career goals?

    _____
    _____
    _____

c. In what areas do you need to grow personally?

d. What are you personally doing to develop others?

7. Analyze your organization's current leadership team. Enter specific data for each member. Use the rating legend to document each member's personal situation pertaining to retirement status, succession planning priority, and the criticality of a replacement moving into their position.

    Rating legend for completing chart on following page:

Retirement status	
Retirement likely within one year	A
Retirement likely within two years	B
Retirement likely within three years	C
Succession planning priority	
High	1
Moderate	2
Low	3
Will not promote further	0
Criticality	
Must hit the ground running	1
Be functional within six months	2

8. Based on the information gleaned through this succession planning analysis, provide an overview of findings.

Position	Incumbent name	Hire date	Promotion to current position	Pensionable years of service	Age	Retirement status	Number of staff on current promotional list	Criticality of position	Number of staff ready now	Number of staff ready in one to two years	Succession planning priority

9. Based on the succession planning analysis, list five immediate action steps that need to be implemented to ensure organizational stability.

   a. _____
   b. _____
   c. _____
   d. _____
   e. _____

CHAPTER 6

EMPLOYEE PERFORMANCE EVALUATIONS

"There is no one here who is qualified to evaluate my performance!" he yelled. The night before, during one of our weekly meetings, I had announced to the volunteer members of Hampton (IL) Fire Rescue that we would begin doing annual performance evaluations for all personnel. Now I was sitting with a friend and fellow department member who, in raised voice and using language that would make a sailor blush, was telling me that no one on our volunteer department was qualified to evaluate him.

He had missed the meeting due to being on shift at his career department and had just listened to the recorded audio from the meeting. He was red-faced, neck-veins-dancing, fist-slamming, spittle-flying furious! During the meeting, I had outlined the reasoning as well as the criteria that would be used and how I felt that it would become an important part of our quest to improve and create organizational excellence. I had also made a comment during the meeting that I didn't care if someone disagreed with this policy; it was going to be implemented anyway. He felt that this comment was pointed directly at him. Truthfully speaking—it was.

He was angry that I would institute such a ridiculous, harebrained policy and accused me of intentionally announcing it at a meeting when he was not present. (He was correct on this last point as well.) He reminded me that he was a professional career firefighter and that he volunteered in our department because he liked the work—but he didn't need to be here. He told me how he had spent years learning and developing his craft, reading everything he could get his hands on, and traveling to several metro departments across the country, on his own dime, to ride and learn from his brother firefighters. He went on to say that I was the one who appointed him as our department's training officer and that it was his job to make sure our personnel were operationally ready. How dare I suggest that anyone he was responsible to train would be able to evaluate his performance—even me.

As much as I would like to say that I handled this interaction with wisdom and the calm demeanor and professionalism of a battle-tested fire chief, I did not. Rather, I handled it like a relatively new and inexperienced fire chief with only a couple of years in the CEO position—which I was.

If I had the chance for a do-over, I would have handled it differently. First, I would work harder to bring along my command team, selling them on the importance of performance evaluations and seeking their input on development of the evaluation criteria. Although we discussed the new policy, I did not do enough to get the buy-in of the majority. I felt strongly about the need for evaluations, so I just forced the issue.

For the record, forcing issues is usually a bad idea. I realize that I may never have won the support of my training officer on this issue, but I could have tried harder. He was a good guy who was not sold on the direction I wanted to head. It would have been much better for us to come to point where we agreed to disagree on this issue than for me to ignore his thoughts and simply push forward.

Secondly, when I announced the new program to the entire membership, I needed to do more to paint a picture of how performance evaluations would help each of us improve and be better, stronger firefighters. Many of the department members felt that this move was making their volunteer job feel more like their real job. They wanted to be their best, but they weren't sure that pointing out their weaknesses was going to serve as a motivational tool to help them improve. Some of their weaknesses were caused by challenges external to the department, like their real job, childcare, and family commitments. They needed help juggling the balls, not someone telling them how many times they dropped one.

Thirdly, I would have not publicly picked a fight with my training officer. That was stupid!

I still believe that it is important to sit down with employees and talk about expectations and goals. But I no longer believe that the historical model of performance evaluations is best. What my volunteers needed back then was help in figuring out how to manage competing demands. They needed department leadership to help them be successful team members. They also needed help and support in figuring out how to reach their personal goals. I don't think the needs of today's employees are any different regardless of whether they are volunteer or paid. I just didn't know back then of a different model.

WHAT IS A PERFORMANCE EVALUATION?

A performance evaluation is a process that uses predetermined standards to evaluate employee performance over an established period. The overarching goal is to provide review and feedback with a focus on strengths, areas of needed improvement, and the establishment of goals. A properly developed and delivered evaluation should be sufficiently specific to inform and guide the employee in their performance of duties (Haigh, 2018).

When it comes to performance evaluations, people typically reflect back to their school days and find the appraisal system reminiscent of that dreaded report card. They may have been out of school for years, but they still have vivid memories of report cards chastising them for their inability to sit quietly, focus for more than five minutes, and not mess with the student sitting at the desk nearby (Haigh, 2018). For many of us, if we brought home a report card with good grades and the teacher said that we generally behaved, there was an associated reward of praise from our parents. If the report card had a few not so good marks, the praise went away, and we were held accountable. Truth be told, the employee performance evaluation has a similar feel. For many, it also contains a built-in carrot-and-stick enticement of a possible increase in pay if you do well. So, in this case, if it looks like a report card, works like a report card, and feels like a report card, it's probably a report card. Remember, we generally don't like report cards!

I will admit though, as my wife reminds me, there are a few individuals who do like report cards. These are the folks who work hard, get the good grades, and check all the boxes on the proverbial list of being a great student. These are the same employees who transition into the workforce and then are angered and demotivated by the boss who does not recognize their hard work. They need the encouragement that comes from words of affirmation and support. As a boss, if you want to make this type of employee feel devalued, fail to prepare a well-thought-out evaluation and then don't take the time to meet and have a meaningful discussion. By failing to take these basic steps, you can easily turn off and hurt a strong supporter of your team.

BOSSES DON'T LIKE PERFORMANCE EVALUATIONS EITHER

Like their employees, most bosses don't care for performance evaluations either. They feel that they are time consuming, uncomfortable, often have questionable value, and sometimes can hurt more than they help. With rare exception, they generally provide no clear road map for direction and are simply a tradition that accomplishes little or nothing in the way of true feedback or help to the employee (Entrepreneur Magazine, 2019).

Many bosses place so little emphasis on the process that they do them quickly to just get them done without providing much thought or insight into what the evaluation says. Sometimes they do nothing more than check the rating categories, and if using an electronically driven evaluation system, let the software autogenerate the narrative statements. This lack of attention makes the employee feel that the boss is just going through the motions and that either the process is not important enough to garner their time and attention or they as an employee are not that important to the boss. Lack of attention to detail in this area sends the wrong message to the employee about both the process and how they are valued by the organization.

To make an annual evaluation work effectively, bosses need to be collecting information and making notes about their employees throughout the review period. These journal notes should capture both the positives and negatives, which then become tools for the boss to use to complete the performance evaluation. These journal entries take time and thought, which is becoming a rare commodity in the fast-paced, understaffed work environment of today. Without these notes, bosses are left to rely on their memories, which will generally produce an evaluation that focuses only on the last few months of the review period simply because they can't remember the rest of the year in any real detail.

BUILDING A CASE AGAINST AN EMPLOYEE

The general premise behind modern employee evaluations is that they are designed to help the employee so that the employee is then better equipped to help the employer. At its most basic level this is a true statement. The flip side of this equation is the fact that evaluations are also a tool that allows employers to document and then notify employees of noted poor performance or areas where they need to improve. This documentation and notification are important steps in the process required to justify future employee discipline or discharge.

When helping an employer to prepare a case to justify employee discipline or even dismissal, one of the first documents that legal counsel will request is past performance evaluations. They want to determine if they can document trends and how these trends link to the reasons behind the proposed discipline. As an employer, did you tell the employee that they were not performing up to an acceptable standard? Did you tell them what they need to do to improve and how long they have to make the corrections? Did you tell them the consequences of not making corrections? When this is all documented within the context of the evaluation, the employer has a much greater chance of winning a court case related to an adverse employment decision.

So, in many ways, the employee has a right to be concerned about the annual performance evaluation because it helps to legally set the stage for the future of their employment. Purists of the process place great emphasis on the evaluations because they see this as a necessary step in maintaining order and discipline within the organization. They need to dot the i's and cross the t's so they are prepared if they need to separate ties in the future.

It could thereby also be argued that managers who place little emphasis on the process are failing both their employees as well as their employer. They are not using the process to help their employees develop so that they can better serve the organization and address any weaknesses while encouraging them to continue building on their strengths. They are also not documenting the truth related to poor performance or areas of weakness, as they see it, that might be needed in the future to help paint a picture of a problem employee.

VARYING PICTURE OF TRUTH

This issue of truth as seen through the eyes of the manager can also be problematic. Managers often bring all kinds of personal quirks and biases, both conscious and unconscious, that influence how they review employees. As an example, I once had a lieutenant who rated a firefighter very harshly compared with how other firefighters were rated. This firefighter for all intents and purposes was an outstanding employee who was a model of excellence. He came to work early, performed exceptionally well whether assigned to a fire apparatus or an ambulance, spent time training new employees, taught in the department's rookie school, and coordinated our hazardous materials team. Most people believed that he was on the fast track to becoming an officer and would be a good one at that. Yet listening to his immediate supervisor and reading his performance evaluations gave a very different impression.

Digging into this situation revealed a couple of things. First, I believe that the firefighter's lieutenant realized his capabilities and was often harder on him than the others. Second, I also think that there was a level of underlying jealousy in that he was rapidly becoming better than his boss. This lieutenant was a quality and dedicated officer, but below the surface I think an unidentified bit of envy existed that caused the firefighter to be rated lower than he deserved.

In the end, this firefighter left our organization to take a career deputy fire chief role in a smaller department, where he rose to the rank of fire chief after a few years. He is now serving as a fire chief back in the Chicagoland area and doing a great job. The evaluation rating of his lieutenant compared against how he was viewed by another organization clearly brings to question a supervisor bias.

Another example of the varying picture of truth based on a supervisor bias involves a 2014 research study of 28 Silicon Valley tech companies. The researcher compared the tone or content of reviews based on employee gender. Using almost 250 performance reviews completed by both male and female managers, three-quarters of the women reviewed were criticized for their personalities using words and phrases such as "abrasive," "bossy," "sharp-elbowed," "territorial," and "not a team player." On the other hand, the men were lauded for their "aggressive" behavior and encouraged to do more of it (Snyder, 2014).

Rater bias is certainly evident in this study of how men and women are expected to behave in the workplace. Also, problematic and consistently evident in performance ratings is the idiosyncratic rater effect. In this phenomenon, the rater varies how much value they personally attach to any given review category based on their own personal feelings. Although we assume that evaluation ratings measure the performance of the ratee, most of what is being measured by the ratings is the unique rating tendencies of the rater. Thus, ratings actually reveal more about the rater than they do about the ratee (Buckingham & Goodall, 2015). As an example, a rater may apply more focus and attention on how a candidate performs tactically at a fire scene than on how they perform on an EMS call. Although both incidents are important to the department's service delivery, the focus on the fire scene with a downplay on patient care tells us a great deal about the rater. Despite this, the employee's direct supervisor remains the one best qualified to rate the employee.

HOW DID WE GET HERE: HISTORY OF PERFORMANCE REVIEWS

So how did we get here, and what do we need to do?

I think that the first step in making a change is to understand the history behind performance evaluations and then take a look at our current workforce and how performance evaluations impact their overall performance.

Performance appraisals trace their history back to the U.S. military's merit rating system created during World War I. This system was designed to identify soldiers who were poor performers and who then became candidates for discharge or transfer. By the time the United States entered World War II, the Army had transitioned this system into a forced ranking model used to identify enlisted soldiers who had the potential to become officers.

Following World War II, U.S. companies saw this rating model as a beneficial system to identify prospects for managerial positions, believing that a strong merit rating would predict good future supervisors and managers. At least initially, this system gave no thought to how it might help improve employee performance, only how it could be used to pick those with the potential to lead. As the years passed, the rating system transitioned into a process to identify strong employees and allocate rewards for good performance (i.e., pay increases). It is estimated that by the 1960s, close to 90% of all U.S. companies utilized some type of merit or performance rating system (Cappelli & Tavis, 2016).

Although tweaks have been made, the employee performance evaluation process used by most employers has changed little and has remained fairly constant over the last 100 years. Although disliked by both employees and managers, the process continues to be utilized with few modifications. A study by the Society for Human Resource Management showed a third of all HR professionals were unsatisfied with their organization's employee appraisal model. They identified deficiencies in areas of leadership development, employee coaching, 360-degree feedback, and development planning (Fandray, 2001).

Even with this acknowledgment by the folks who are tasked with managing the process, the performance evaluation model continues to be part of the foundation of how we evaluate our employees.

EVALUATION PROCESS AND THE CHANGING GENERATIONS

As the evaluation process has moved forward over the years, there has been little attention given to the impact of changes within the generational workforce. I evaluated my 20-something firefighters today in the same way that I was evaluated 30 plus years ago when I entered the fire service. Thirty years ago, I was evaluated the same way firefighters were evaluated in the 1960s. Nothing significant has changed in our evaluation models, yet we have had sweeping changes in how we operate, the skills and services we provide, the tools we use, and the educational base of our employees.

Using evaluation tools developed for the traditionalists or silent generation (those born before 1945) on the Generation Z (mid-2000s to 2015) workforce is a system destined for failure. Over the last 100 years the impact of computerization, the internet, globalization, communication, the financial revolution (cash to plastic), the generalized U.S. shift from manufacturing to services, and women entering the workforce have all modified how we do business. Yet we continue to push forward in many ways with an evaluation system that closely resembles its origin and one that can be argued is failing to accomplish our end goal of improving employee performance.

MILLENNIALS

It is estimated that millennials compose 50% of the U.S. workforce and will reach 75% by 2025. Millennials were born sometime in the early 1980s to mid-2000s. They grew up in an electronics-filled, online, and social network world (Rouse, 2015). In many departments, millennial employees are beginning to reach the tenure to test for and be promoted into first-line supervisory positions. In other cases, they are the firefighters with 5 to 10 years on the job who are today's workhorse employees who are riding the ambulances, pulling hose, searching buildings, washing rigs, and swirling toilets. In order to figure out how to properly evaluate these team members, we need to better understand how they work and think.

Workplace satisfaction has been identified as a top priority for millennials, followed closely by work-life balance. Making money is not the driving motivational factor. For millennials, doing work that is meaningful often outpaces the drive for wealth. However, finding stable employment to pay for higher education and soaring housing costs makes competitive pay a priority when they look for a job. They search for employers who provide a career path that supports their lifestyle outside of work while still offering competitive pay, a positive career trajectory, and overall balance (Kohll, 2018). The fire service is very attractive to millennials based on these factors.

In general, millennials are team players and feedback seekers. They prefer feedback in real time with frequent and instant reviews. Remember, this generation has grown up with social media and the instant feedback provided through a tweet or a post. They like to check in often with their bosses to receive encouragement and tips on how to improve their performance. They are eager to learn and develop and flexible to change when provided with clear and specific direction through honest dialogue. They appreciate a heads-up on their behavior, skills, job knowledge, and attitude. They highly desire a mentor-type relationship where they receive guidance on their individual careers rather than the typical dictatorial officer relationship that is common in the fire service (Hernandez, 2015).

This need for real-time feedback is not compatible with our historical once-a-year performance review systems. In a 2015 TriNet survey of 1,000

U.S. full-time millennial employees, more than half felt that their managers were usually unprepared, and nearly 9 out of 10 said that they would feel more confident if they could have more frequent performance conversations with their managers. Failure of employers to meet the feedback need of these employees breeds job dissatisfaction and will often result in high employee turnover. In the survey, more than 28% of respondents said that they had reacted to an annual performance review by looking for a new job (TriNet, 2015). This contrasts with the fact that satisfied millennials are known for their loyalty and often become huge employee advocates for their respective organizations. Part of their satisfaction is driven by timely and appropriate feedback (Hernandez, 2015).

GENERATION Z

Generation Z individuals were born in the mid-2000s to 2015. Like millennials, for Generation Z, technology shapes almost every aspect of their lives. Gen Z workers prefer face-to-face communications with their coworkers and managers rather than an email or phone call—especially when it involves talking about their careers. Supervisors who seek to avoid tough conversations and who are not prepared to fully engage related to performance and mentorship will have a hard time with these employees.

Added at the top of their list of desires found in the perfect job is the ability to be flexible. The stringent paramilitary schedule of the fire service workday simply does not make sense to them. They believe that since they are here for 24-hours, if the rigs are ready to go and they answer their calls, the rest of the stuff can be done anytime throughout the shift. They ask questions that frustrate their bosses and challenge the long-held traditions of the fire service:

> Lieutenant—would it be okay if we do housework today from 1500 to 1630 instead of 0930 to 1100?

They are not moved by the fact that for the last 50 years the department has done housework at 0930. For the officer who believes the world will stop spinning on its normal axis if the toilets do not get cleaned at 0930, they should expect to have a hard time supervising this up-and-coming workgroup.

Gen Z employees come to work and find no motivation or passion in a clock-watching, "look busy" mentality that does not allow for overall flexibility. Officer statements of "This is how it has always been done" without strong supporting reasons make no sense to them and lead to workplace dissatisfaction.

This flexible work pattern goes along with their mentality of less segregation between work and life. They are much more about balancing and making the two seamless, so work gets done anywhere, anytime, without sacrificing either one. Leaders of this next generation should expect these employees to be willing to take on a special department project and to do it in their off-duty time but request to be allowed to work remotely instead of coming into the firehouse or office. This will allow them to get the work completed while still maintaining off-duty balance. As an example, they may do the work from home and list the time on their timesheet while still being able to take care of their children while their spouse is at work. They will see this as a win for all involved. For the traditional manager, this type of flexibility may be beyond their comfort level, resulting in frustration on both the part of the manager as well as the employee.

In general, Gen Z employees, although young, are willing to work hard because they have set clear career paths for themselves. They want to work with coworkers who work as hard as they do and in workplaces where they are able and encouraged to do their best work (Florentine, 2018).

GENERATION X AND BABY BOOMERS

These two groups represent our incumbent or more senior employees. The boomers are those approaching or already in the 60-plus-year-old category. In paid departments, these employees are rapidly headed toward retirement, yet for the volunteer/paid-on-call agencies, they continue to be a mainstay of many organizational operations. The Gen Xers are in their late 30s to mid-50s and are in many cases the current leaders managing our departments/districts. In 2018 it was estimated that 51% of all the world's organizations were being led by Gen Xers (Moran, 2018).

Both of these two groups can be management challenges in that they view life very differently from each other as well from our millennials and Gen Z employees. The boomers are frequently seen as workaholics and lovers of overtime. They value hard work and financial security, and they are less inclined to embrace change. Conversely, the Gen Xers are the children of the boomers who came home from school each day to an empty household due to both parents working. This produced a generation of workers who are independent, self-reliant, self-sufficient, resourceful, and adaptable. They tend to despise micromanagement and, based on their upbringing as latchkey kids, want balanced lives with family-friendly workplaces. Due to their independent nature, they often dislike the mentor style and frequent feedback-driven leadership styles craved by the millennials and Gen Z employees. They see this as micromanagement and would much rather be left alone to figure out solutions and do the work.

For managers, having four distinctly different generations of workers working side by side in our firehouses can be both a tremendous opportunity as well as a tremendous challenge. Implementing a one-size-fits-all approach to performance evaluations simply will not work. The differences in backgrounds, experiences, culture, and beliefs necessitate creative and divergent processes to both lead and evaluate these workers.

CHANGING THE PROCESS OF EVALUATIONS

Dr. Tori Culbertson of Kansas State University and colleagues wrote,

> It is not so much that the performance review needs to be abolished, but we need to fix what is broken. Instead of limiting ourselves to formal performance appraisals conducted once or twice a year, we need to think about performance management as a system that is linked with the strategy of the entire organization.... If we are going to have once-a-year evaluations, we shouldn't expect them to work. (Culbertson et al., 2013)

Deloitte is a UK-incorporated multinational professional services firm that has spent several years assessing its employee performance evaluation system and has found its model to be increasingly out of step with the company's objectives related to training, promotion, and pay.

As part of its redesign process, Deloitte came to realize that once-a-year reviews did not fit within a "real-time world," and conversations about year-end ratings were less valuable than conversations conducted in the moment about actual performance.

So how do you build a "real-time" evaluation system? Deloitte started the challenging process by first looking closely to see if there were identifiable similar performance characteristics from its company's best performing teams. (Remember, the fire service operates based on teams.) What became evident is that the teams were all very much "strengths oriented." Meaning, members are placed in positions where they can employ their personal individual strengths on a regular basis to consistently benefit the team.

So how can you best evaluate this team process? Deloitte found that a briefing-style evaluation process based on the teamwork needed to complete a project rather than a singular focused evaluation of a single employee's performance worked best. When digging into its model, the process is very similar to what the fire service does during a postincident analysis following a major emergency response.

Next, the company focused on the idiosyncratic rater effect by changing the questions asked as part of the performance review process. Although supervisors rate other people's skills inconsistently, they are highly consistent when rating their own feelings and intentions. Deloitte leveraged this by developing a system that does not ask about the skills of the employee but rather about the supervisor's own desired future actions with respect to the employee (Buckingham & Goodall, 2015).

- How strongly would I push to always have this person on my team?
- Using a scale of 1 to 5, at what level do I believe this person is at risk for low performance? What could I do to make this person a stronger team member?
- When coaching this team member—how am I doing?
    - Regular interaction, review, and feedback?
    - Do I focus regularly on what is the simplest view of the employee or on what is the richest?
    - How do I personally help the employee be their absolute best?

Like Deloitte, Adobe has also significantly modified its employee appraisal system by implementing a more informal check-in process that takes place throughout the year, with employees receiving feedback on what they're working on at any given moment. The theory is that the more often employees receive feedback, the more productive and focused they'll be (Entrepreneur Magazine, 2019). Additionally, these check-in meetings are an excellent time to discuss and establish some individual employee performance goals or benchmarks. These can then become discussion points for subsequent meetings to help the employee remain focused on achievement of the agreed-upon goals.

MAKING IT APPLICABLE TO THE FIRE SERVICE

A strength of the fire service work environment is the fact that both supervisors and subordinates live, eat, and bunk together in the same firehouse. Company officers ride the rig with their personnel, and frequent communication takes place throughout the shift regarding a variety of topics. This strength is also a weakness in that this closeness often becomes a barrier for effective discussions related to job performance. Although bosses frequently talk with their members, the focus on being the team's coach is often missing (Haigh, 2018). In the words of the legendary football coach Tom Landry, "A coach is someone who tells you what you don't want to hear, who has you see what you don't want to see, so you can be who you have always known you could be" (Landry, n.d.). Coaching the team, including the individual players, is not a once-a-year event. It involves constant day-to-day listening, motivating, training, inspiring, and mentoring with a focus on building a team that can consistently carry out the mission of the department.

To do this, bosses need to go beyond the normal day-to-day interactions and hold regular check-in meetings or "coaching sessions" with subordinates. These meetings should be used to not only check on project status and performance but also to evaluate the emotional and psychological health of their team members. As fire service leaders, we too often find ourselves asking our people, "How is it going?" and we receive the standard answer of "okay." This is because firefighters by nature are self-reliant, mission-driven personalities who simply do the work and fix the problems. They will complain about the evening dinner and how the prior shift did not empty the kitchen garbage but will typically not tell the boss about a problem until you are behind the proverbial eight ball (Haigh, 2018).

To minimize this situation, during the check-in meetings the use of a topic template is often helpful for both the firefighter and the supervisor. These templates give the firefighter time to think, prepare, and focus on the discussion while giving the supervisor a forum to ask questions, make notes, and provide feedback. The topics may be as broad as the following:

- What are you working on?
- What are you thinking about?
- What are you learning?
- What are your needs?
- How can I help?

In essence, they are asking about employee goals and how the employee is working to carry out the organization's mission and values. They are also asking about personal goals and how they can assist in helping the employee achieve something important that is important to them, even if it is not work related. I personally like to start my employee review/coaching meetings with the statement, "Take me through what I need to know." This is my time to LISTEN and to let my team member speak. There is plenty of time for follow-up after you learn where they are at and what they want to discuss. These check-in meetings can be thought of as a personalized postincident analysis or debriefing. Also, these check-in meetings serve to break review periods down into sprints rather than extended year-long time frames. This gives all involved the opportunity to remember what happened and to discuss actions in a timely and much more agile fashion. Also, discussions need to be more holistic and focus on goals and strengths, not just past performance. To make this work, a once-a-year evaluation discussion process is not even close to being what is needed (Cappelli & Tavis, 2016; Murphy, 2005).

DROPPING THE ANNUAL PERFORMANCE REVIEW

It is not just the fire service that is looking for something different. Other government organizations like NASA and the FBI are also beginning to rethink their approach to annual performance appraisals. These agencies are recognizing the need for supervisors to do a better job coaching and developing their subordinates (Cappelli & Tavis, 2016). In their article "The Performance Management Revolution," Peter Cappelli and Anna Tavis list three main reasons that companies are dropping their annual performance appraisals for a more frequent check-in process:

- People development
    - Firms are doubling down on development, often by putting their employees (who are deeply motivated by the potential for learning and advancement) in charge of their own growth. This approach requires rich feedback from supervisors—a need that's better met by frequent, informal check-ins than by annual reviews.
    - Keeping good people is critical. Companies are trying to eliminate dissatisfiers that drive employees away. Annual reviews are on the list, since the process is so widely reviled and the focus on numerical ratings interferes with the learning that people want and need to do. Replacing this system helps managers do a better job of coaching and allowing subordinates to process and apply the advice they are receiving.
- The need for agility
    - Projects are short-term and tend to change along the way, so employees' goals and tasks cannot be plotted out a year in advance with any real accuracy.

- General Electric (GE) has implemented touchpoints based on two simple questions:
    - What am I doing that I should keep doing?
    - And what am I doing that I should change?
- The centrality of teamwork
    - This teamwork mindset more closely follows the natural cycle of work.
    - Conversations between managers and employees should occur when projects finish, milestones are reached, challenges pop up, and so forth. This allows people to solve problems and develop skills for the future.

CASE STUDY: H.P.F.D.

Over the last couple of years, as an experiment, we implemented regular check-in meetings at Hanover Park Fire Department. In the first year, officers were instructed to begin scheduling bimonthly meetings with their full-time personnel and were instructed to ask the following four questions:

- How are you doing?
- How are we (the department) helping you to meet your goals?
- In your opinion, are you meeting the department's goals?
- How can we (the department) help you?

The answers to these questions were documented in the "Journal" section of the village's electronic employee performance review system. During each subsequent meeting, the same questions would be reviewed, along with any associated progress, changes, or challenges. This repeated review kept issues at the forefront of the discussions and ultimately allowed many items to be addressed.

As time passed, officers began to learn of several private issues that were impacting employee performance. Issues with a spouse, kids, parents, finances, off-duty destructive behavior, and a myriad of other issues all came to light and became part of the discussions. Officers were instructed to hold these issues in confidence (unless it involved suicidal concerns or illegal activities), to help where possible, and to use maximum discretion in documenting some of these issues within the journal system. The goal was to encourage officers to serve as confidants and coaches and to not use what they were learning in a way that could be used against the employee in the future. Understanding that this is a thin line, I am proud of how my team handled these situations.

In year two of the experiment, we changed up the questions and expanded our evaluator group. The new questions were designed to address information learned in year one:

- On a scale of 1 to 10, how would you rate your current level of reliability/dependability?
- What challenges exist that impact your reliability/dependability?
- How can we (the department) help eliminate/manage these challenges?
- What are two job-related goals that you would like to accomplish in the next 12 months?

With this change we also added our part-time firefighters to the bimonthly check-in process. In order to keep this manageable for our officers, we decided to train our acting company officers (ACOs) as evaluators. In Hanover Park, ACOs are certified at a minimum to the Fire Officer I level and currently hold an active position on the lieutenants' promotional eligibility list. For this process, each part-time firefighter is assigned to one of these ACOs. The ACOs use the same questions and conduct the check-in meetings in the same way as the fully commissioned officers.

Overall, I believe the process has worked well. As with anything, it has been a learning process, and we have had a few officers and ACOs who need to be periodically reminded to do the documentation. They are good at holding the check-in meetings but not so good at writing them down. We have also had a few challenges where subordinates have indicated a goal or a desire, for example to attend a training class, which was talked about during the meetings but never relayed to the training officer so something could be scheduled. This moving of pertinent information up the chain of command continues to be an evolving process. I am confident that they will get there.

CHECK-IN MEETINGS FOR BOSSES

As members climb the fire service organizational ladder and are assigned increased responsibilities over divisions and major projects, providing written weekly/monthly reports to your upper-level chiefs should be used to document accomplishments, concerns, and project status. These reports then become the basis for the check-in meetings with the upper-level chief officers. These reports should become part of the employee's personnel file. When the reports are viewed collectively, they should provide a clear visual progression on project management, organizational challenges and accomplishments. It is a good idea for upper-level chief officers to review, at least quarterly, with their employee the reports in a "collective format" to piece together the big picture and to identify trends—both positive and negative (Haigh, 2018).

This system also works well for the fire chief/CEO related to their direct report. Figure 6–1 is an example showing part of a weekly update that I submitted to the village manager related to fire department operations. Attached to this is information related to call volume, data from special studies, and correspondence such as thank you letters or information from our partner agencies. Basically, anything that I wanted to document and share with the village manager got placed on this report. Then when we met together, she used the report to drive our discussions.

GOAL FOCUS

Regardless of your supervisory level within the organization, when meeting with subordinates, discussions need to focus on goals: goals for the organization as well as the employee's personal goals.

These check-in meetings, coupled with some basic documentation, form the basis for a performance management system that replaces the annual performance evaluation model. The big annual meeting is gone, replaced by a much more relevant, real-time review that focuses on helping the employee be successful rather than reviewing issues that are in the past. These check-in meetings are much nimbler, less stressful, and typically quicker, and they do a better job in providing performance reviews to the employee as well as ensuring that goals are kept as a priority for both the boss and the subordinate.

DOCUMENTATION

It has been a long-standing management principle that the annual performance evaluation serves as a tool to document ongoing performance problems, thereby lending legal support for employee discipline. As bosses, we generally dislike always thinking about the need to prepare a case for employee discipline, yet documentation remains as a cornerstone component in the management of any serious disciplinary action. Bottom line—even the best legal team cannot help in managing your personnel issues if you don't have a paper trail (Haigh, 2018).

This is where the regular check-in meeting excels. During these meetings, the supervisor is constantly working through the process of making employees aware of their performance strengths and deficiencies. Once aware of concerns,

> **WEEKLY UPDATE – FIRE DEPARTMENT**
> **JULY 31, 2020**
>
> ➢ **ADMININISTRATION**
>   o Chief Haigh represented the department at the second townhall meeting of "Peace Together our Community"
>   o Battalion Chief Jasper represented the fire department at the monthly MABAS Division 12 meeting.
>   o Chief Haigh, HR Director Kurcz, and Assistant Chief Fors represented the Village at the quarterly Fire Department Pension Board Meeting.
>
> ➢ **TRAINING DIVISION**
>   o A representative from Columbia Southern University held three meetings offered to all FD personnel to discuss options for members to continue furthering their formalized education. The department is a registered learning partner with Columbia Southern University.
>   o The department's training division is working to update our Job Performance Requirements (JPRs) including all times associated with select skills. Drills were held to validate times as part of this review process.
>
> ➢ **EMERGENCY MANAGEMENT DIVISION**
>   o Assistant Chief Fors held the Village PPE group meeting.
>
> ➢ **INSPECTIONAL SERVICES DIVISION**
>   o Three solar permits were issued during this period. The total solar permits issued in 2020 is 49, and 154 total solar permits have been issued for 2018, 2019, and 2020.
>   o Verandah - all units have been issued certificates of occupancy except for two units that need to replace floors where the wood has warped.
>   o Habitat for Humanity's project on Greenbrook Court is progressing. Foundations have been completed for the third duplex and they are ready for steel. The pond in functioning and blanket and seed have been installed.
>   o Menards - 7435 Barrington Road has installed seed and blanket, trees have been installed and the seating area has been completed.
>   o The division has seen a significant increase in submittals for and issuance of permits along with an increase in phone calls and emails to answer questions.
>   o Inspector Bertolami has kept in touch with restaurants and businesses to discuss COVID-19 compliance along with his normal inspections.
>   o San Marcos Mexican Grille located at 1222 Lake Street has opened.

Figure 6–1. An update from the fire department to the village manager.

the employee can make corrections, and the supervisor has a means to check on progress and lend support where possible. Supervisors need to document these discussions and be sure to make the employees aware of the consequences of future infractions or their failure to improve performance (Haigh, 2018).

In order to maintain the paper trail, it is best to utilize a standardized departmental system where notes and discussion templates can be maintained for future reference, retrieval, and review by other department supervisors. Many departments have an electronic journal built into their performance appraisal software. Using this journal tool is an excellent way to accomplish this documentation goal. For those who don't have access to such a system, an easy and fast model that works well is to have a conversation with the employee, then follow up with an email or memo memorializing the discussion. Then simply save and file the email or memo.

SOME EMPLOYERS STILL REQUIRE ANNUAL REVIEWS

Many employers (e.g., municipalities, counties, townships, departments, districts) still require their supervisors to conduct annual performance evaluations regardless of the information available to show how ineffective they truly are. Bottom line: change takes time. In these cases, I encourage fire chiefs/CEOs to go ahead and implement the check-in meeting process with their employees and then use the data and documentation to complete the required annual reviews. Depending on your system, you may be able to easily paste the information from your check-in meetings into the narrative sections of the annual evaluation tool and then calculate the numerical ratings based on the information you have been collecting over the past 12 months.

CALCULATING PAY RAISES

Pay raises in many career departments are determined by the collective bargaining process, and evaluations have little impact on the amount of raise an employee receives. Other collective bargaining agreements (CBA) reference evaluations and adjust pay based on performance. Members not covered by a CBA almost always receive pay raises based on their annual review. The challenge of doing away with the annual review along with the typical numerical rating system makes pay raises more subjective and less programmed. It is also hugely difficult to have a serious, open discussion about problems while also hitting the employee with the negative consequences of a low pay increase (Cappelli & Tavis, 2016). This model clearly moves the evaluation process away from performance improvement and employee development to a clear carrot-and-stick process that is probably not healthy for the employee, the supervisor/manager, or the organization.

When moving to the check-in meeting process, the focus needs to transition to accomplishment of goals and how the employee handles the ever-occurring hurdles that seemingly always pop up throughout the course of any given year.

Employees should be evaluated and pay raises awarded based on how they work through the problems and how they manage goals. Pay raises given using this criterion cause the employee to develop a clearer focus on performance and to move away from simply putting in their time.

ONE SIZE DOES NOT FIT ALL

Although check-in meetings offer a distinct advantage over the annual performance evaluation model, evaluation components can and should vary depending on what you are trying to measure and the varying needs of the department at any given time. Fire departments are not static machines and have different performance evaluation needs at different times. Utilizing the same evaluation system year after year is likely to not be in the best interest of the organization and will frequently become stale and not produce the desired results. Using different measurement tools at different times is often helpful in uncovering issues and finding hidden strengths and weaknesses that can be used to enhance employee performance (Haigh, 2018).

A few different examples of evaluation tools designed to fit varying situations and organizational needs are detailed next.

JOINT SUPERVISOR EVALUATIONS

In some cases, fire department employees are supervised by more than one equally ranked supervisor. This is often the case with volunteer and part-time firefighters as well as firefighters who are routinely moved between houses. To provide a comprehensive evaluation, feedback is required from each of the supervisors. A joint supervisor evaluation form works well to elicit feedback from each supervisor.

The form I used captures both a rating score as well as individual comments from each of the supervisors. The tool asks each officer to numerically rate the firefighter on a scale of 1 to 4 using a variety of performance competencies. Under each competency, they were asked to also provide a brief comment justifying their rating. The evaluations are then collected so one single document can be created. By using this joint evaluation process, it allows each officer/ACO that the firefighter works with to provide individual feedback while having all the information compiled into a single, easy-to-understand rating tool.

One officer or ACO is designated as the lead supervisor who is responsible to collect the evaluations, compile the finalized document, and then do the employee review/check-in meeting. They take the scores and narratives from the officers/ACOs and average the numbers for each competency to determine a collective score for the section. They then cut and paste the narrative statements into the comments sections. To calculate the final score for the entire evaluation, they again average the scores from all competency areas to determine the finalized rating score. Here is an example of the form used by Hanover Park (fig. 6–2).

SELF-EVALUATION

A self-evaluation tool is often useful to help employees explore their personal strengths and weaknesses. No employee wants to be a problem or weak link within an organization. Regardless of the attitude they project, deep down they want to do a good job. The self-evaluation tool helps employees to critically analyze their own performance and to provide insight into their strengths and weaknesses. It is also recognized that most people are consistently interested in themselves—their own insights, achievements, and impacts. A self-evaluation system provides a tool for employees to share what is best about themselves and how their strengths can positively impact the department (Buckingham & Goodall, 2015).

I first began using self-evaluation tools while serving as fire chief in King (NC). The tool was very straightforward and allowed both the employee and the supervisor to complete a side-by-side rating on a variety of broad categories. This side-by-side rating was designed to be a discussion starter and comparison between what the employee believed compared with the supervisor's perspective. It also asked questions related to areas in which the member was involved as well as how we could better use their unique skills. The process was simple and effective and was used by both our career officers as well as our volunteer officers to evaluate their personnel.

In recent times, I have slightly reworked King's self-evaluation tool for use during my end of employee probation meetings. It is my practice to personally meet in a one-on-one meeting with all department members at the end of their new hire probationary period or their probationary period that follows a promotion. These meetings were designed to talk about what they learned, how they have grown and developed, and their goals for the future. I was specifically interested in how they think they are doing. I was also keenly interested in their goals, and we usually spend most of the time talking about what they would like to accomplish and how the department could assist them.

Just prior to beginning the meeting I gave them a one-page self-evaluation form and ask them to take a few minutes to complete it. I told l them to not overthink the questions and that I am most interested in their immediate thoughts. They usually took somewhere around 10 minutes to complete the

HANOVER PARK FIRE DEPARTMENT
Part-time Firefighter Performance Evaluation

Employee Name:
Job Title: Part-time Firefighter/EMT/Paramedic
Review Period Start:
Reviewer:

RATINGS:
Low = Needs Improvement
          Meets job requirements
          Effective Performance
High = Outstanding Performance

Performance Competencies

	Low 1	2	3	High 4
Productivity	☐	☐	☐	☐

Comments:

	Low 1	2	3	High 4
Respectful Workplace/Cooperativeness	☐	☐	☐	☐

Comments:

	Low 1	2	3	High 4
Dependability/Reliability	☐	☐	☐	☐

Comments:

	Low 1	2	3	High 4
Job Knowledge	☐	☐	☐	☐

Comments:

Figure 6–2a. A joint supervisor evaluation form, page 1 (courtesy of Hanover Park Fire Department).

	Low			High
	1	2	3	4
Safety Awareness	☐	☐	☐	☐

Comments:

	Low			High
	1	2	3	4
Customer Relations	☐	☐	☐	☐

Comments:

	Low			High
	1	2	3	4
Planning/Organization	☐	☐	☐	☐

Comments:

	Low			High
	1	2	3	4
Overall	☐	☐	☐	☐

Comments:

Goals

Must establish two goals for completion in 12-month period from time of evaluation.
1)

2)

_____   _____
Lieutenant   *Date*

Figure 6–2b. A joint supervisor evaluation form, page 2 (courtesy of Hanover Park Fire Department).

document, and then we meet. Keep in mind that I had already talked with their field training officers (FTOs) and supervisors and had read their formalized evaluation documents. This gave me insight to compare their self-evaluation ratings with what I had already learned. Since beginning this practice, I have been amazed at the clarity of self-awareness displayed by the members, and I typically found their thoughts are very much in line with what I have found during my preparation discussions prior to the meeting. The tool gives a great launching point to begin discussions.

A perception fallacy about self-evaluation tools is that if you ask an employee to rate themselves, they will only give themselves high marks since no one is going to tell their boss where they are truly weak. This has not been my experience. Employees tend to be brutally honest in self-evaluations. Using this tool, which relies on the idiosyncratic rater effect, can offer a unique picture of the employee whereby the supervisor can learn a great deal and will typically gain information that can be used as discussion starters during the regular check-in meetings.

Figure 6–3 shows the evaluation document I used.

A word of caution in using self-evaluation tools: Occasionally you will find an employee who has a genuine lack of perception related to their performance. They simply do not see their performance in the same light as their supervisors. In these cases, the employee genuinely believes that they are doing a great job, which contrasts with what the organization and their supervisors see. These situations can become challenging and frustrating for all involved. My best advice is to openly discuss the different perspective with the employee in an attempt to help them see and understand where the discrepancies exist. I have found that using specific examples often helps, but more frequently than not, I leave these meetings fully convinced that the employee really does not understand the problem. Sometimes I will even bring in a second supervisor in hopes that the employee will hear the same message in a different way and will then understand. Sometimes it works, and sometimes it does not. In the end, both you and the employee are frustrated, which is not the desired outcome of these review/check-in meetings, but it is also a reality in managing people.

360° EVALUATION

360° evaluations are utilized primarily for supervisors. Think of them in a similar fashion to getting a 360° at an emergency scene. Experienced officers will tell you that what is seen from the street side may be very different from what is happening in the rear. The 360° evaluation tool helps to provide the

HANOVER PARK FIRE DEPARTMENT

EVALUATION FORM

NAME_____

DATE_____

CATEGORIES	5	4	3	2	1	COMMENTS
Conflict Resolution						
Commitment to Quality						
Communication						
Dependability						
Emergency Response						
Initiative						
Job Knowledge						
Physical Condition						
Teamwork						
Use of Technology						
Leadership						
Impact and Influence						

Rating Scale: 5 – Outstanding 4 – Good 3 – Satisfactory 2 – Needs Improvement 1 – Poor

AVERAGE SCORE: [ ] **DIVIDE BY THE NUMBER OF CATEGORIES**

FINAL SCORE: [ ]

In what areas of the department are you currently involved?
_____

How do you use your unique skill sets to impact the department?
_____

How can your abilities and skills be better utilized to impact the department?
_____

Figure 6–3. Evaluation form from Hanover Park Fire Department.

supervisor with the full picture of how their leadership style and decisions are impacting the people they lead or work with.

The 360° evaluation is designed to seek input from all involved pertinent parties. It starts with the supervisor being rated, moves to gain the input of subordinates or coworkers, and then ultimately the information gleaned coming back to the one being rated in a comparative format.

To accomplish this, the one being rated is given the 360° evaluation tool and asked to rate themselves in a variety of categories. Likewise, several of their subordinates and coworkers are given the same rating tool and asked to evaluate the individual in the same categories. The evaluations are then collected and compared with each other by a third party. A finalized report is generated that shows the details of the comparison and the revealed gaps in perception. This analysis, coupled with some significant self-reflection, is designed to help the one being evaluated to see their performance through the eyes of their team (Haigh, 2016). The steps involved in the process are detailed as follows:

- Implementation: How the 360° process is implemented varies from organization to organization and from individual to individual based on their overall sphere of influence. In a small 30- to 40-member volunteer/paid-on-call department in which the governing board is using the tool to evaluate the performance of the fire chief, they may have all members of the organization complete the evaluation tool.

    Another department may break the evaluation process into work groups. As an example, a battalion chief managing a shift of 15 to 20 personnel could easily have all members complete the evaluation, excluding those members not assigned to their shift.

    The process could be modified for a division chief who has 75 to 80 members who indirectly report to them and who are within their sphere of influence. In this case, it might be much more efficient and accurate to obtain a sampling that includes battalion chiefs, the division's administrative assistant, and select officers and firefighters who interact regularly with the chief.

    Or in another case, a large department with multiple battalion chiefs working each day may want to do a peer 360° to determine the perspective of coworkers as it relates to effective daily interaction toward meeting the mission of the department. When it comes to selection of the evaluation group, there is no specific answer. Each department needs to handle the selection of participants on

a case-by-case basis, carefully considering what makes sense for the organization and the information you are looking to obtain (Haigh, 2016).
- Confidentiality: The key to gaining meaningful information from those who are doing the rating is confidentiality and trust. All parties need to be encouraged to share their opinions openly and honestly without fear of retaliation. To accomplish this, the process needs to be a blind data collection process.

    As an example, when I am conducting a 360° evaluation, I send all raters the 360° evaluation form electronically; they complete the document, print it out, seal it in an unmarked envelope, and turn it in to a designated individual. This individual is responsible for collection but also maintains a signature sheet whereby all assigned evaluators are tracked to ensure completion of the 360°. The one being evaluated completes the same 360° evaluation form except that theirs is marked with their name or some other type of designated tracking method. An internet-based survey system such as SurveyMonkey can be used to efficiently accomplish the same goal. The importance is not in how the data is collected but in the quality of the data received. The only two true essential aspects are the following:

    - The process is mandatory and needs to involve each of the assigned raters.
    - The process needs to be confidential.

- Question development: When developing the tool, emphasis needs to be placed on what is being assessed and the questions you want answered. The tool also needs to be concise yet thorough, and easy to complete without being overly lengthy. A tool I have used successfully is modeled after one borrowed from private industry (ODW Logistics, 2010).

    When developing the assessment questions, I typically pull together a group of supervisors who will be evaluated using the tool. We talk about what information they would find helpful to know. We then turn this information into questions asked on the 360° evaluation tool. This process allows the tool to obtain information that will be immediately helpful to the supervisor. Once the tool is developed, it is sent to the employees. The employees are given specific instructions on how to complete the tool and how it should be submitted. Figure 6–4 shows the tool Hanover Park is currently using to evaluate battalion chiefs.

FIRE DEPARTMENT 360 FEEDBACK TOOL
Battalion Chief _____ - Evaluation Period 2020

CHARACTERISTIC	FACTORS OF PERFORMANCE	RATING					COMMENT
		Poor	Fair	Average	Good	Excellent	
Fire Department Leadership							
Question 1	Communicates a clear and compelling vision for the fire department.	1	2	3	4	5	
Question 2	Possesses knowledge of the fire service including area and national trends.	1	2	3	4	5	
Question 3	Develops ideas and strategies that create value and vision.	1	2	3	4	5	
Question 4	Is well versed in department culture and history.	1	2	3	4	5	
Question 5	Maintains a thoroughly professional and quality driven environment and culture.	1	2	3	4	5	
Question 6	Keeps others focused on the vision and strategic goals of the department.	1	2	3	4	5	
Question 7	Respected by others as being knowledgable in the field and among their peers.	1	2	3	4	5	
Leading Change							
Question 1	Champions and paves the way for positive change.	1	2	3	4	5	
Question 2	Anticipates key changes affecting the department and the fire service.	1	2	3	4	5	
Question 3	Identifies obstacles before they become a crisis.	1	2	3	4	5	
Question 4	Challenges conventional practices in search of new/effective solutions.	1	2	3	4	5	
Question 5	Inspires creativity in others by being open minded to new ideas.	1	2	3	4	5	
Customer Focus (Internal and External)							
Question 1	Understands what staff wants from their department.	1	2	3	4	5	
Question 2	Consistently advocates for quality.	1	2	3	4	5	
Question 3	Anticipates needs and requirements and proactively works to present solutions.	1	2	3	4	5	
Question 4	Is resilient and committed to find solutions with even the most demanding customer needs.	1	2	3	4	5	
Question 5	Staff feels supported and respected.	1	2	3	4	5	
Question 6	Represents the department well in meetings and communicates with residents and businesses.	1	2	3	4	5	
Results Orientation							
Question 1	Consistently delivers superior results and can be counted on.	1	2	3	4	5	
Question 2	Empowers others with necessary resources, authority and responsibility.	1	2	3	4	5	
Question 3	Follows up and monitors situations to proactively develop solutions.	1	2	3	4	5	
Integrity and Trust							
Question 1	Sets a good example and models department values.	1	2	3	4	5	
Question 2	Is direct, straightforward, and honest in dealings.	1	2	3	4	5	
Question 3	Accepts responsibility for problems and does not blame others.	1	2	3	4	5	
Question 4	Insures credit for "job well done" is given to appropriate parties.	1	2	3	4	5	
Question 5	Can be trusted to represent my interests, even if I am not around.	1	2	3	4	5	
Communication							
Question 1	Delivers a clear, concise, and timely message.	1	2	3	4	5	
Question 2	Adapts well in presenting information to different audiences and situations.	1	2	3	4	5	
Question 3	Effectively communicates through all levels of the organization and shares information.	1	2	3	4	5	
Question 4	Makes it safe for others to express concerns/ideas and listens openly.	1	2	3	4	5	
Question 5	Actively asks for feedback and advice from others.	1	2	3	4	5	
Planning and Organization							
Question 1	Keeps people focused on the department's key initiatives and priorities.	1	2	3	4	5	
Question 2	Uses time effectively and accommodates workload as required.	1	2	3	4	5	
Question 3	Develops plans that are realistic and effective in meeting departmental objectives.	1	2	3	4	5	
Decision Making							
Question 1	Is not afraid to make tough decisions in a timely manner.	1	2	3	4	5	
Question 2	Demonstrates good judgment, common sense, and gathers appropriate data when making decisions.	1	2	3	4	5	
Question 3	Ability to efficiently and effectively identify a problem and quickly execute a solution.	1	2	3	4	5	
Question 4	Enables and supports others in their decision making process.	1	2	3	4	5	
Passion / Commitment to Excellence							
Question 1	Demonstrates work habits which exceed expectations.	1	2	3	4	5	
Question 2	Exhibits values which promotes and supports a commitment to excellence.	1	2	3	4	5	
Question 3	Supports others in their quest for continuous personal improvement and development including coaching and mentoring.	1	2	3	4	5	
Functional Excellence (Emergency Operations)							
Question 1	Provides clear radio communication.	1	2	3	4	5	
Question 2	Provides direct and concise orders.	1	2	3	4	5	
Question 3	Enables and supports the proper use of the chain of command.	1	2	3	4	5	
Question 4	Identifies and verbalizes potential hazards.	1	2	3	4	5	
Question 5	Uses critical incidents as learning tools.	1	2	3	4	5	
Question 6	Recognizes emotional inconsistencies after critical calls.	1	2	3	4	5	
Question 7	Provides critical debriefing immediately after incident.	1	2	3	4	5	
Overall							
Question 1	Overall, I rate Chief _____ leadership as _____	1	2	3	4	5	
Question 2	Overall, I rate Chief _____ relationships with others as _____	1	2	3	4	5	
Question 3	Overall, I rate Chief _____ contributions and achievements as _____	1	2	3	4	5	
Question 4	Overall, I am "this excited" to be on Chief _____ team _____	1	2	3	4	5	
Question 5	Overall, I respect and trust Chief _____ "this much" _____	1	2	3	4	5	
GENERAL COMMENTARY							

Figure 6–4. Feedback tool (courtesy of Hanover Park Fire Department).

- Data analysis: Once all evaluations are collected, the data analysis process begins. I have found that utilization of an Excel spreadsheet works very well. The goal is to use the data to create a comparison line chart. The line chart visually displays the gap between performance perceptions (see fig. 6–5).

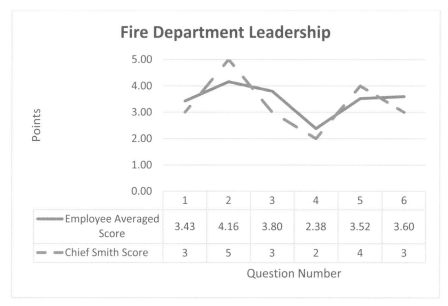

Figure 6–5. Ratings chart for leadership questions. The numbers 1 through 6 on the top row of the table correspond to the questions asked. In this case there were six questions related to leadership, as listed before.

In a perfect world, there would be no difference between the two lines. In the real world however, a gap almost always exists. The 360° tool is used to help discover that gap so actions can be taken to correct or minimize the distance between the two perspectives. The analysis is not intended to make conclusions on who is right or wrong but to identify perceptions that can be discussed and actions that can be taken to improve performance where necessary (Haigh, 2016).

CASE STUDY EXAMPLE

To help in understanding the process, I created a hypothetical evaluation using the generic "Chief Smith" as our evaluated candidate. Figure 6–5 and 6–6 show the gap rating (i.e., comparison line chart) for each question asked using categories of *Fire Department Leadership* and *Leading Change*. I then also created counseling charts based on the divergence of scores (tables 6–1 and 6–2).

Here are the questions asked in the *Leadership* category:

1. Communicates a clear and compelling vision for the fire department.
2. Possesses knowledge of the fire service, including area and national trends.
3. Develops ideas and strategies that create value and vision.

4. Maintains a thoroughly professional and quality-driven environment and culture.
5. Keeps others focused on the vision and strategic goals of the department.
6. Respected by others as being knowledgeable in the field and among their peers.

Analyses of this hypothetical scoring category suggest that both Chief Smith as well as his subordinates rated his performance in the ranges of average (3) to good (4). This might suggest that his performance is acceptable but not in the "high performer" category. A closer look might suggest the following action items from the divergences noted (I typically only address with candidates the ones with divergences of near or greater than one full point).

Questions asked in the *Leading Change* category:

1. Champions and paves the way for positive change.
2. Anticipates key changes affecting the department and the fire service.
3. Identifies obstacles before they become a crisis.
4. Challenges conventional practices in search of new/effective solutions.
5. Inspires creativity in others by being open-minded to new ideas.

Comments provided by the evaluators on the 360° evaluation tool should also be shared with Chief Smith. These can be delivered in a variety of ways

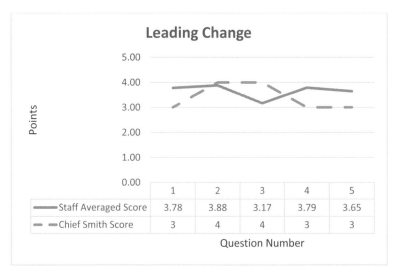

Figure 6–6. Ratings chart for leading change questions. The numbers 1 through 5 on the top row of the table correspond to the questions asked related to leading change.

but, if possible, should be supplied word for word. Looking at them unchanged can help Chief Smith better gauge some of the feelings and perceptions (right or wrong) of his people. Armed with this knowledge, he can make course corrections to his leadership style that will make him significantly more effective.

The 360° evaluation tool is a versatile instrument that can be designed and modified to create a learning tool to assist fire service leaders at all levels. Once the process is completed, the subject evaluated is armed with a powerful tool to help them reflect on their individual leadership style and how it is

Table 6–1. Questions and suggestions for improvement.

Question number	Question	Areas of concern	Suggestions for improvement
1	Communicates a clear and compelling vision for the fire department.	Score within standards	Continue doing what you are doing.
2	Possesses knowledge of the fire service, including area and national trends.	Chief Smith believes that this is a personal strength with subordinates scoring at a lesser level.	Work to better share Chief Smith's knowledge in this area with subordinates. This might include more focused attention on communicating with personnel.
3	Develops ideas and strategies that create value and vision.	Chief Smith rated himself lower in this category than his subordinates.	Continue working in this area since success is being realized based on the analysis of subordinates (i.e., "keep doing what you are doing").
4	Maintains a thoroughly professional and quality-driven environment and culture.	Chief Smith appears to be weak in this area.	This is an area of importance. Chief Smith should focus specific attention here.
5	Keeps others focused on the vision and strategic goals of the department.	Score within standards	Continue doing what you are doing.
6	Respected by others as being knowledgeable in the field and among their peers.	Score within standards	Continue doing what you are doing.

Table 6–2. More questions and suggestions for improvement.

Question number	Question	Areas of concern	Suggestions for improvement
1	Champions and paves the way for positive change.	Chief Smith rated himself lower in this category than his subordinates.	Communication with subordinates is needed to determine why this disparity exists.
2	Anticipates key changes affecting the department and the fire service.	Score within standards	Continue doing what you are doing.
3	Identifies obstacles before they become a crisis.	There appears to be concerns in this area.	Analyze areas where decision making has been slow and problems have escalated or had potential to escalate.
4	Challenges conventional practices in search of new/effective solutions.	Chief Smith rated himself lower in this category than his subordinates.	Communication with subordinates and self-reflection of past decision making is needed to determine why this disparity exists.
5	Inspires others creativity by being open-minded to new ideas.	Chief Smith rated himself lower in this category than his subordinates.	Communication with subordinates to determine what he is doing right and how he can continue and bolster his performance in this area.

impacting those who work with them. Based on the information gleaned, coupled with self-reflection, this allows the individual to adjust as necessary to do their absolute best. It is also important as a supervisor to realize that some of the staff will not like you and will make harsh comments (the reality of being a manager), while others will love you and will make overly positive comments. As you reflect on the comments you most likely will need to adjust your style to somewhere in the middle (Haigh, 2013).

As a point to ponder, something that I have noted over the years when it comes to using 360° evaluation tools is that the supervisors who typically have the highest ratings are the ones who are strong, out-front leaders who daily embody and demand excellence in themselves and those they supervise.

Leaders who try to lead their teams based on a laissez-faire attitude of letting things take their own course typically are scored poorly. Supervisors who are mean and manage by fear and intimidation are scored the worst.

BIG PICTURE

Employee performance reviews are important. More frequent and timely discussions help to maintain employee focus and effectiveness. If you plan to make changes to your evaluation process, do it slowly. Take time to listen, test, and refine the process to ensure that it best meets the needs of your specific organization and employees. What works for one department may not be a good fit for another. It is okay to be creative if the focus remains on improving employee performance. As a manager, begin viewing yourself as the team's coach. As a coach, your job is to produce a winning team, and for your team, lives literally hang in the balance.

REFERENCES

Buckingham, M., & Goodall, A. (2015, April). Reinventing Performance Management. *Harvard Business Review, 93*(4), 40–50. http://hbr.org/2015/04/reinventing-performance-management.

Cappelli, P., & Tavis, A. (2016). The Performance Management Revolution. *Harvard Business Review, 94*(10), 58–67. https://hbr.org/2016/10/the-performance-management-revolution.

Culbertson, S. S., Henning, J. B., & Payne, S. C. (2013). Performance Appraisal Satisfaction: The Role of Feedback and Goal Orientation. *Journal of Personnel Psychology, 12*(4), 189–195.

Entrepreneur Magazine. (2019). *Performance Reviews*. https://www.entrepreneur.com/encyclopedia/performance-reviews.

Fandray, D. (2001). The New Thinking in Performance Appraisals. *Workforce*, 36–40.

Florentine, S. (2018, June 20). Everything You Need to Know About Generation Z. *CIO*. https://www.cio.com/article/3282415/everything-you-need-to-know-about-generation-z.html.

Haigh, C. A. (2013). *Administrator Evaluation Process*. Roselle, IL: Medinah Christian School.

Haigh, C. A. (2016, July). The 360° Performance Evaluation Tool. *Fire Engineering*. https://www.fireengineering.com/leadership/the-360-performance-evaluation-tool/.

Haigh, C. A. (2018, August). Rethinking the Annual Performance Evaluation. *Fire Engineering*. https://www.fireengineering.com/leadership/rethinking-the-annual-performance-evaluation/.

Hernandez, R. (2015, November 3). Here's What Millennials Want from Their Performance Reviews. *Fast Company.* https://www.fastcompany.com/3052988/heres-what-millennials-want-from-their-performance-reviews.

Kohll, A. (2018, March 27). The Evolving Definition of Work-Life Balance. *Forbes.* http://www.forbes.com/sites/alankohll/2018/03/27/the-evolving-definition-of-work-life-balance/#4ca986ab9ed3.

Landry, T. (n.d.). *Runner Things #2684.* www.fuelrunning.com/quotes/.

Moran, G. (2018, April 19). Why You Need to Pay Attention to Gen X Leaders. *Fast Company.* https://www.fastcompany.com/40558008/why-you-need-to-pay-attention-to-gen-x-leaders.

Murphy, J. D. (2005). Flawless Execution: *Use the Techniques and Systems of America's Fighter Pilots to Perform at Your Peak and Win the Battles of the Business World.* New York: HarperCollins.

ODW Logistics. (2010). *360 Evaluation Tool.* Columbus, OH.

Rouse, M. (2015). *Millennials (Generation Y).* TechTarget. http://whatis.techtarget.com/definition/millennials-millennial-generation.

Snyder, K. (2014, August 26). The Abrasiveness Trap: High-Achieving Men and Women Are Described Differently in Reviews. *Fortune.* http://fortune.com/2014/08/26/performance-review-gender-bias/.

TriNet. (2015). *Performance Reviews Drive One in Four Millennials to Search for a New Job or Call in Sick.* https://www.trinet.com/about-us/news-press/press-releases/survey-performance-reviews-drive-one-in-four-millennials-to-search-for-a-new-job-or-call-in-sick.

QUESTIONS FOR FURTHER RESEARCH, THOUGHT, AND DISCUSSION

1. Conduct an analysis of your organization's workforce based on age. How does your workforce break down by generational categories both in number and percentage?

   a. Boomers (1955–1964):_____
      Percentage of workforce: _____%
   b. Generation X (1965–1980):_____
      Percentage of workforce: _____%
   c. Millennials (1981–2001):_____
      Percentage of workforce: _____%
   d. Generation Z (2002–2015):_____
      Percentage of workforce: _____%

2. Where within your organization do you see the potential for supervisor (rater) bias or a negative idiosyncratic rater effect?

   _____
   _____
   _____
   _____
   _____

   a. How is rater bias or a negative idiosyncratic rater effect applicable to you personally as an evaluator/supervisor? How does it impact how you are evaluating your employees?

   _____
   _____
   _____
   _____

   b. What steps can you take to minimize this situation in how you personally evaluate employees?

   _____
   _____
   _____
   _____
   _____

3. Develop five "topic template" questions to be used when meeting with subordinates. Questions should revolve around the status of assigned projects, overall employee performance, and an assessment of the employee's current emotional and psychological health.

   a. _____
   b. _____
   c. _____
   d. _____
   e. _____

4. Based on the varying types of employee evaluation tools that exist, if you were asked to recommend changes to your organization's overall evaluation process, which tools would you recommend and why?

   _____
   _____
   _____
   _____
   _____

5. In developing a 360° evaluation tool that would allow feedback on your personal performance as a fire officer/supervisor, provide five categories where you would like feedback, including corresponding questions for use within each category.

   a. Category: _____

      i. Question: _____
      ii. Question: _____
      iii. Question: _____

   b. Category: _____

      i. Question: _____
      ii. Question: _____
      iii. Question: _____

   c. Category: _____

      i. Question: _____
      ii. Question: _____
      iii. Question: _____

d. Category: _____

   i. Question: _____
   ii. Question: _____
   iii. Question: _____

  e. Category: _____

   i. Question: _____
   ii. Question: _____
   iii. Question: _____

6. What roadblocks and hurdles exist within your own organization related to changing or modifying your current employee review process? Identify the challenges and list action steps required to begin discussions and actions designed to facilitate change in this area.

  a. Challenge: _____

   i. Action step: _____
   ii. Action step: _____
   iii. Action step: _____

  b. Challenge: _____

   i. Action step: _____
   ii. Action step: _____
   iii. Action step: _____

  c. Challenge: _____

   i. Action step: _____
   ii. Action step: _____
   iii. Action step: _____

  d. Challenge: _____

   i. Action step: _____
   ii. Action step: _____
   iii. Action step: _____

  e. Challenge: _____

   i. Action step: _____
   ii. Action step: _____
   iii. Action step: _____

CHAPTER 7

CONDUCTING INTERNAL INVESTIGATIONS AND EMPLOYEE DISCIPLINE

It's early on a Sunday morning in December 2015, and I have just brewed my first cup of coffee. The house is quiet, except for the pitter patter of dog paws crossing the kitchen floor looking to steal a quick treat before heading out for her morning romp in the snow-covered backyard. It has been a busy few weeks with numerous evening budget meetings and a teaching gig the weekend before. As I reflect on the budget meetings, even this morning I am concerned about whether my overtime budget allotment will cover what looks to be a couple of long-term firefighter vacancies due to injuries.

Firefighter injuries are the great nemesis of the fire service. Injuries cause vacancies, and almost all vacancies result in overtime. Overtime eats away at the dollars needed to purchase equipment, train personnel, and buy the supplies needed to keep the firehouses operating and rigs going up and down the streets. Although it may not have yet been said by anyone in village hall, any experienced fire chief knows that reductions in company staffing levels are always on the table by managers and boards when overtime costs begin to exceed what can be absorbed into an already tight budget. Experienced fire chiefs also know that reductions in staffing lead to more injuries, allowing the cycle to quickly spiral out of control.

As my morning unfolds, I am just looking for a few peaceful minutes to myself before the family awakes and the Sunday morning routine begins. The holiday season is in full bloom with the Christmas tree decorated with cherished heirlooms and twinkling lights. The fragrance of the coffee in my cup is helping to ease me into the day. Christmas is just ahead, and I am looking forward to spending time with extended family members and parents that seem to be aging faster each day. Just a few minutes of quiet and a good cup of coffee, and I will be ready!

But not today.... My cell phone begins to ring, and I see by the caller ID that it is the union president. It is 7:32 a.m., and I instinctively know that my peaceful morning is coming to an abrupt end. Before I answer the call, a second call pops up on the screen indicating my assistant fire chief/executive officer. Simultaneous calls from both my number two and the union president can only mean one thing: all is not well in the Hooterville Fire Department.

I answer the union president's call first—always best to keep him as an ally instead of an enemy. He leads with the words "...We have a problem." He tells me how he has just been notified that one of our up-and-coming firefighters, one of our rock stars, a firefighter whom I know personally and hold in very high regard, has suffered a terrible lapse in judgment and was arrested in the early morning hours for driving under the influence. He goes on to tell me that the situation gets worse in that the firefighter allegedly attempted to use her position to garner leniency from the arresting officer, and when this didn't work, the firefighter, who is a person of color, then started calling the officer a "racist cop." He wants to know what I plan to do.

After thanking him for his call and reiterating how much I value our relationship, I click off and return the call to my assistant chief/executive officer. The chief answers my call on the first ring and leads with those now familiar words "...We have a problem." I tell him that I just spoke with the union president and that I know the basics of the situation. The chief is an experienced and battle-tested administrator who back in the day also served as union president. He knows the players and the rules and has already done his homework to discover the preliminary facts of the case.

He confirms the story relayed by the union president and then begins to explain what we know. He tells me that the firefighter was stopped after midnight, arrested, and taken to jail. The firefighter was slated to be on duty at 0700 hours this Sunday morning and was a no-call/no-show. When this firefighter did not show up for work, the concerned crew began making calls to family and friends. After reaching a family member and having a quick conversation about the firefighter's whereabouts, the department then received a call from the firefighter. The firefighter spoke to the on-duty battalion chief and requested 12 hours of sick leave. When pressed about what was going on, the firefighter admitted to the arrest and again requested sick leave. The battalion chief told the firefighter to not report to duty and that he would be notifying the bosses. The assistant chief, like the union president, was now asking how I would like to proceed.

How would I like to proceed? Frankly, I would like to start this day over, ignore the two phone calls I just received, and go back to the peaceful tranquility of the quiet morning that I had planned. But I know that this is not an option. I am disappointed, saddened, and a little hurt that this firefighter would place me and the department in this situation. I realize that this is not a situation that I can fix and one that will likely result in a negative outcome for both the firefighter and the department. There will be no winners here . . . only losers.

I know that I need to call the village manager as well as the mayor. The last thing I need to have happen is for the mayor to get a call before he is briefed and have this thing get to the news media. He is going to be unhappy and will probably tell me that he wants the firefighter discharged immediately and how this is an unacceptable action by one of his people. I realize that I can probably calm him down and convince him to let us investigate further, but I also know that this situation and my actions will end up being judged well outside the walls of the department. An entire troop of "Monday-morning quarterbacks" will be standing by to review my every move. As I mentally steady myself to make the needed phone calls, I remind myself that this is all part of the weight and loneliness of the fifth bugle.

WHERE TO BEGIN

No matter the quality of an organization's employees and the strength of its supervisors, sometimes things happen. Employees make mistakes, have lapses in judgment, and do things inconsistent with the high standards of public service. When this occurs, you as the fire chief/CEO will most assuredly be pressed by all sides to make a quick and decisive decision. In my experience, this is the worst thing you can do. Quick decisions do not give you time to think, fact find, and get advice from wise counsel. Quick decisions lead to mistakes, which tend to be costly for both the employee and the organization. This is not to say that discipline should not be timely—it should. It just should not be rushed to the point where you make a mistake. Calm, reasoned, and well-thought-out action always wins the day.

STOPPING THE CLOCK

Sometimes though, based on the seriousness of the alleged offense, a level of immediate action is required. In these cases, you need to "stop the clock" so you can figure out how best to proceed. This stopping of the clock sends a message to both the accused as well as the onlookers that you are taking the situation seriously. What I define as "stopping the clock" is placing the member on paid administrative leave. This move is in my opinion the safest and best first action. It allows you to restrict the employee from the work of the organization while not causing an adverse employment decision that will likely be a point of litigation in the future.

Administrative leave is a temporary restriction from a job assignment while maintaining the employee's pay and benefits. When placing an employee on this type of leave, the wording most often used in the employee's official letter is as follows:

> You are being placed on paid administrative leave pending the outcome of an internal investigation.

By taking this action, you are giving yourself the time you need to begin the process of figuring out what happened and determining your next steps. It puts the employee on notice that the leave is not open-ended and will conclude upon completion of the investigation. It also signals that an action will result from the investigation (exoneration, discipline, termination). As a policy, when I place someone on administrative leave, I make them surrender their badge and credentials. This prevents them from officially acting on behalf or representing the organization during the period of the investigation. It also shows the seriousness of the situation.

If you lead a volunteer department, the action of administrative leave is the same, except that the issue of compensation does not apply. If, however, you pay benefits, provide insurance policies, or have a retirement fund for volunteers, you still need to make any contributions on their behalf just as if they were working. For volunteers, I not only make them surrender their credentials, but I also make them remove or cover any response lights installed in their private vehicles.

Lastly, this step of administrative leave helps you as fire chief/CEO address what I call the "Front Page of the Chicago Tribune Test." As the leader of an emergency service organization, the last thing you or your organization needs is for the media to get wind that something serious has happened and that you have taken no action. In the court of public opinion as told on the front page of the local paper (or social media), you as the leader need to act. Without a level of action, it looks as though you are either ignoring the situation or preparing to cover it up. You need to send the message that something happened, you know about it, you are investigating, and further action is forthcoming.

Unfortunately, it does not always make the story go away, but it sends the right initial message.

INTERNAL INVESTIGATIONS

Internal investigations are used to uncover the facts to determine whether misconduct occurred. Once you have placed an employee on paid administrative leave and said that you are going to investigate, internals are the investigative tool used. You need to enter these investigations recognizing that you rarely have all the facts. An internal investigation, when conducted appropriately, works to sort out fact from fiction.

WHERE WE STRUGGLE

In the local fire department, unless you are large enough to have a division of professional standards (a.k.a. internal affairs), internal investigations are an area where many of us routinely fall short. Most of us have not been trained in how to properly conduct an internal investigation, including how to properly interview those involved. Our partners in law enforcement excel in interviewing and knowing how to read responses to questions, body language, and the overall demeanor of the one being questioned. They quickly get a feel as to whether someone is being forthcoming or hiding information. They also know how and when to turn up the heat or ease off on the tension. Police officers literally live and die based on their ability to read people. This is simply not the case for those of us in the fire service, and therefore we tend to be weak when it comes to investigative questioning.

Added to this lack of knowledge and experience is the fact that we often go into investigations with preconceived opinions. We know our personnel, we see them at their best and their worst, we know where they struggle, and we know their families, work habits, and histories. I would argue that we often are too close to our members, especially in small organizations, to be able to fully separate these factors and perform an unbiased investigation. These factors can cloud the investigative picture and sometimes prevent us from doing the right thing related to employee discipline. I have come to believe that in cases involving serious allegations, it is almost always best to ask for assistance from outside.

In some communities the human resource department has trained investigators who can conduct or assist with these investigations. In Hanover Park I often used our contract labor attorneys to help with internal investigations. They have a team of experts who both understand how to conduct investigations as well as knowledge of the legal aspects required to protect both the rights of the organization and the employee. I am also aware of a couple of private investigator firms who specialize in this type of work and are available for hire on a contract consulting basis. A few departments/districts utilize their communities' law enforcement agency, especially if the police agency has a dedicated internal affairs division. In others, members who are sworn as both firefighters and police officers and who are assigned to their fire investigation units become the ones responsible for conducting internal investigations.

I do not know that any one model is superior to another, other than for the fact that an organization needs to know before they have a problem how they are going to handle these situations. Sticking your head in the sand and pretending that they are not going to happen is simply naïve and will end up catching you unprepared. My best advice: figure out what you are going to do, create an internal investigation policy, and when something goes wrong, follow the policy.

ROLE OF THE FIRE CHIEF/CEO

The fire chief/CEO should not be the one responsible for conducting or managing the internal investigation. This allows the fire chief to remain objective and neutral to the facts of the case since they will be the decision maker for the final disposition of the findings of the investigation. This is often challenging for two reasons:

1. Some departments/districts have built a culture where the fire chief/CEO is responsible for all discipline. The culture says that the chief will investigate and determine what is going to happen. All other members take a "wait and see approach," letting the full weight of the situation be shouldered by the fire chief/CEO. If this is the model used within your organization, I would strongly recommend that you rethink this process. I would argue that divesting control of the investigation into the hands of others while maintaining control of the final decision is the best model.
2. The public, when making a complaint, always wants to talk to the fire chief/CEO. They think they have a right to go directly to the top of the organization, and they want you to personally hear their grievances. When you hand them off to someone responsible for internal investigations, they are often offended and think that you are not taking their complaint seriously. They miss the fact that by having them talk to an investigator you are saying that you *are* taking their story seriously

and that you want to make sure that you organizationally get it right. I always try to explain this as I hand them off. Sometimes they understand, and sometimes they do not. Regardless, this does not mean that it is not the right thing to do. Keep in mind that you as fire chief/CEO have a responsibility to the public, but you also have a responsibility to your employees and the organization. To support all involved effectively, you need to remain impartial so you can make the best decision based on facts.

ORGANIZATIONAL CULTURE AND INTERNAL INVESTIGATIONS

Department/district personnel need to be taught that they have both the authority and the responsibility to bring forward a situation that has the potential to negatively impact the public, the members, or the organization. This understanding of personal responsibility does not manifest itself once a situation has occurred but rather is a cultural value that is built into the fabric of the organization.

I know of some departments/districts that have a culture that says, regardless of the situation, we take care of our own. This means that lying and covering up issues to protect fellow employees is culturally expected. Anyone not adhering to this way of thinking is ostracized and pushed from the fold. Other organizations have a more cutthroat mentality where team members lay in wait to find someone making a mistake so they can jump and attack. There is no comradery or support for each other but rather everyone is out for themselves and willing to step on one another to continue moving ahead.

Neither of these scenarios are desirable. Members need to support and care for each other while also setting a high standard of expectations. The culture of an organization is the key to having few disciplinary problems and being able to effectively manage them when they occur. The culture is "the way we do things here" and subsequently sets the standard for what is acceptable behavior and how those who choose to deviate from the culture are held accountable. This standard of expectation becomes that little voice in the back of our heads guiding us on how to make decisions that are acceptable to the culture. It helps employees color inside the lines and often solves problems before they occur. Development of this culture does not come overnight and is not driven completely by one person. It comes from the collective whole of the body. When it comes to internal investigations, this culture can make them extremely difficult and painful or swift and efficient.

EMPLOYEE MISCONDUCT

It all begins here. . . . Whether big or small, on duty or off, paid or volunteer, long tenured or just hired, the reason you are conducting an internal investigation is because something happened due to an employee's actions. Think of employee misconduct as an act or omission that if proven true would result in some form of discipline, sanction, or remediation. This might include the following:

- Commission of a criminal act
- Neglect of duty
- Violation of a policy, rule, regulation, or order (Varone, 2014)

COMPLAINT

The allegation of employee misconduct starts with a complaint. The complaint can come from inside or outside the organization. Although called a complaint, it really is the term used to describe a situation where information has been obtained related to an employee's conduct. The complaint may be brought by an individual, or it may be a situation that has occurred where, without an internal investigation, the identity of those involved is not known.

As an example, I recently dealt with a situation (complaint) that involved a case of trespassing perpetrated by one employee against another whereby the one entered the personal locker of the other without permission. We did not have a complaint per se (the owner of the locker did not know that it had been entered), but the department had evidence to believe that an unauthorized entry had occurred. This required the launching of an internal investigation to determine not only the perpetrator but why they would enter someone's locker without authorization.

Organizations need to have set policies for how complaints are to be handled. This includes how complaints are received. Complaints should be accepted in person or via telephone, email, written letter/document, court document, text message, social media, or similar. Organizations should not restrict how complaints are received, thereby allowing the process of submittal to be as broadly construed as possible.

This policy should also include the process for self-initiated complaints based on knowledge gained by leadership. A good example is the unauthorized locker incident I described earlier. In this case we had a vigilant member who identified what he thought was a problem. He brought it to leadership, and based on his information we decided to open an investigation to determine whether his suspicions were correct. Unfortunately, they were. When documenting a complaint, as much detail as possible should be obtained. Information related to dates, times, involved parties, witnesses, and contact information should all be collected. This is all information that the investigating officer will need to begin conducting the internal investigation.

The *Complaint Form* allows documentation of this basic information, including a written summary of the complaint. It also denotes who took the complaint and where the complaint was transferred within the department's chain of command (fig. 7–1).

Since the fire chief/CEO should not be responsible for managing the investigation, an investigating officer needs to be appointed very early in the process. The investigating officer is responsible to the fire chief/CEO and the governing authority for the processing of the complaint, managing the fact-finding investigation, and reporting such for review and action.

INVESTIGATIVE PROCESS: PRELIMINARY INVESTIGATION

The investigative process begins with a preliminary investigation. The preliminary investigation is intended to capture and preserve evidence, identify witnesses, and prepare a record for use later in the investigation. It is not the final investigation, nor should the officer conducting the preliminary investigation pursue matters to all possible ends. This phase is for the sole purpose of freeze-framing as much of the incident as possible (Varone, 2014).

This preliminary phase of the investigation should be conducted and documented by the ranking officer/acting officer initially notified of the complaint. It is their job to determine the exact nature of the situation and obtain information and evidence. Action items include the following:

- Record pertinent contact information for involved parties.
- List involved department/district personnel
- Obtain photographs to document what happened (e.g., of damaged property).
- Order blood/chemical testing of employees (e.g., drug and alcohol testing postaccident).
- If injuries occurred, after medical treatment is provided, obtain photos of the injuries.
- Begin collecting and preserving documentation, radio recordings, dispatch recordings, videos, and so forth.
- Ensure that personnel complete all incident reports per department/district standards (Varone, 2014).

```
┌─────────────────────────────────────────────────────────────┐
│  [FIRE         HANOVER PARK FIRE DEPARTMENT                 │
│  HANOVER                                                    │
│   PARK          COMPLAINT FORM                              │
│   badge]                                                    │
│                                                             │
│  Complainant Name: _____  │
│  Complainant Contact Information:                           │
│      Phone Number: _____   │
│      Address: _____    │
│      Email: _____                    │
│  Date of Occurrence _____  Incident No. _____   │
│  Summary of Incident:                                       │
│  _____  │
│  _____  │
│  _____  │
│  _____  │
│  _____  │
│  _____  │
│  _____  │
│  _____  │
│  _____  │
│  _____  │
│                                                             │
│  Involved Party(s) _____    │
│  Witness(es) _____    │
│  Name of employee receiving complaint _____    │
│  Signature of employee receiving complaint _____ Date__   │
│  Forwarded to Assistant Chief _____ Date__   │
│  Received by Assistant Chief _____ Date__   │
└─────────────────────────────────────────────────────────────┘
```

Figure 7–1. Complaint form used in Hanover Park (courtesy of Hanover Park Fire Department).

Here is a hypothetical example of a preliminary investigation (maybe not so hypothetical).

Following a response to a reported structure fire in a residential neighborhood, your on-duty battalion chief is approached by a neighbor who says that your ladder truck struck and damaged her mailbox. The resident is very agitated and wants to vent about how the firefighters should be more careful. The battalion chief checks the mailbox and finds that it is indeed damaged. He notes that it has red paint transfer onto both the post and the box itself. He then checks the ladder truck and finds scuff marks on the passenger side rear compartment at a level consistent with the height of the mailbox. When he talks to the driver and officer of the ladder truck, neither indicate any knowledge of hitting the mailbox.

As the ranking officer on-scene, the battalion chief needs to begin the preliminary investigation. First and foremost, he needs to remember that he is representing both the organization and you as the fire chief/CEO. He also needs to keep in mind that sometimes things happen when you drive large apparatus to emergency incidents. Considering all these factors, his first action needs to be to de-escalate the emotion of the situation. He needs to calm the complainant and assure her that he is taking the situation seriously. He should not, however, have the firefighter and officer who were involved in operating the ladder truck talk to her. In my experience this serves no good purpose and tends to only escalate the event.

Next, he needs to approach the investigation understanding that he is not here to determine fault but rather to obtain information that will be passed to the investigating officer. To do this, he needs to do the following and document it on the department's/district's complaint form:

- Obtain the name, address, and contact information of the complainant.
- Determine if there were any witnesses, and if so, obtain their contact information as well as statements about what they saw.
- Check to see if any homes or businesses in the general area might have outside surveillance systems that could have caught the accident on video. If so, request a copy.
- Document the names of the employees assigned to the ladder truck.
- Obtain photographs of the mailbox and the surrounding street area (e.g., parked cars, roadwork, barricades, other apparatus).
- Obtain photographs of the damage to the ladder truck.
- Replace the driver of ladder truck and order him to report immediately to your department's occupational health provider for drug and alcohol testing. Have another department member drive him to the provider. Do not let him drive himself.
- Collect all incident documentation and request all radio recordings from dispatch.

- Check to see if any in-car law enforcement video exists, and if so, request that it be archived.
- Have the ladder truck driver, the officer, passengers in the vehicle, and any staff who were in a position to possibly witness the accident complete sworn statements.
- Make notifications up the chain of command so an investigating officer can be assigned.

SWORN STATEMENTS

An important step in any investigation is obtaining sworn statements from the employees involved. A sworn statement is a written document prepared by an employee detailing their knowledge of the facts of a given situation. They are not to include opinions and should only contain factual information related to what was seen, done, and said. Employees are compelled to write them when directed as part of an investigation. Sworn statements are to be considered the same as if taken under oath. Any statements later proven to be an intentional lie will be considered perjury resulting in serious discipline and possible dismissal from employment.

Sworn statements should be taken as soon as possible following the event (fig. 7–2). The more time that passes, the more opportunity there is for memories to be impacted and clouded by firehouse chatter related to the event. To obtain these personal accounts, I use a fillable and expandable electronic form that allows employees to type their statements. Once completed, they should be compiled and transferred to the assigned investigating officer.

Sworn statements are not designed to "get an employee jammed up" but rather to get the employee to document what happened. Depending on the situation, this documentation may however be an admission of guilt that could ultimately result in discipline. Based on this, an employee has the right to delay an order to provide a written statement until they can receive guidance from their union representatives or legal counsel (see the later section in this chapter on employee rights).

INVESTIGATIVE PROCESS: INVESTIGATION

The preliminary investigation should have served to freeze-frame the event so the investigating officer does not lose evidence or witness statements. This gives the investigating officer a basis to begin the next phase of the process. This next phase is designed to review and evaluate what happened and what

Chapter 7 Conducting Internal Investigations and Employee Discipline   **197**

**HANOVER PARK
FIRE DEPARTMENT**

SWORN EMPLOYEE STATEMENT

Employee Name		I.D.	Date:	
Subject				

I, _____, state under oath, under penalty of perjury, that the information above is true and complete to the best of my information and belief.

_____   _____
Signature   Date

Figure 7–2. A sworn statement form (courtesy of Hanover Park Fire Department).

needs to be done. As the assigned investigator, a couple of key questions to ask as you begin work on this phase of the process include the following:

- What are the specific allegations?
- Based on the allegations, does the investigation need to be narrowed, expanded, or dropped?

- Questioning
    - Who has been questioned?
    - What do the sworn statements reveal?
    - Based on the statements and any interviews conducted, where is follow up required?
    - Who still needs to be questioned?
    - In what order should people be questioned?
    - What questions need to be asked?
    - What is the availability of the witnesses you need to question?
- Evidence
    - What evidence exists?
    - What additional evidence would be helpful?
    - Is any evidence missing?
- Are there discrepancies between the evidence and the statements?

In Hanover Park, as part of the internal investigation policy, they have developed check sheets designed to help investigators not miss critical items (fig. 7–3). The *Investigative Case Control Sheet* provides a listing of required documents needed for most internal investigations. Not every investigation will require all of these, but the list serves as a reminder to consider what might be needed. It also provides space to add additional items specific to the case.

The *Investigative Case Activity Log* serves as your ICS 214 log related to the investigation (fig. 7–4). It logs notable activities, contacts, and information as well as dates and times. It documents the investigators' actions taken related to the investigation.

It is always important to remember that there is a strong likelihood that the investigator may be called to defend an investigation in court. Documentation is the key to remembering their actions and findings. These two forms were designed to help document the process.

DISPOSITION OF THE CASE

Depending on what the investigator finds, they have a couple of options related to disposition of the case. The option that should probably be used the most is to simply send the case back to the direct supervisor of the member(s) involved and allow them to provide discipline per department policy. Situations involving employee performance, rudeness, insubordination, minor property damage accidents, damage to tools, and so forth should all be handled by the direct

HANOVER PARK FIRE DEPARTMENT

INVESTIGATIVE CASE CONTROL SHEET

Document Obtained	Date Obtained	N/A to the Case	Required Document
			Complaint Form
			Preliminary Investigative Summary
			Sworn Statements
			Non-agency Witness Statements
			Fire / EMS Incident Reports
			Insurance Reports
			Communication Records
			Lab Analysis Reports
			Medical Records (releases, medical interpretations, etc.)
			Photographs
			Diagrams
			Video Recordings
			Audio Recordings
			Records Checks (criminal, motor vehicle, etc.)
			Law Enforcement Reports
			Media Reports
			Applicable Call Rosters
			Applicable Training Documents
			Equipment Records
			Investigator's Chronology
			Investigator's Handwritten Notes

Figure 7–3. An investigative case control sheet (courtesy of Hanover Park Fire Department).

supervisor. Rarely should these issues get elevated above that level. As Chief Joe Heim of the East Dubuque (IL) Fire Department says, "In these cases, make the officers be officers."

HANOVER PARK FIRE DEPARTMENT

INVESTIGATIVE CASE ACTIVITY LOG

Date	Activity / Contact / Information	Time

Figure 7–4. Investigative case activity log (courtesy of Hanover Park Fire Department).

If the case appears to be of a more serious nature and not something that can be sent back to the immediate supervisor, the investigator will need to conduct the investigation and likely provide a report of findings along with a recommendation to the fire chief. In these cases, the investigator should not hesitate to reach out for assistance when necessary. Bringing in an outside investigator to assist in reviewing the case and providing insight and direction to avoid making mistakes can be critically important. How investigations are conducted has a direct linkage to employee rights. It is easy to make a mistake in this area that will not only jeopardize the investigation but can leave the investigator and the department/district open to grievances and legal action. I would say the most important skill of a good investigator is to know when to ask for help!

Upon completion of the investigation, a written report needs to be provided to the fire chief/CEO. This report needs to include a list of organizational rules, regulations, polices, or standards violated, including any criminal violations uncovered during the investigation. Sometimes the fire chief/CEO may ask for the investigator to also make a recommendation related to discipline or further disposition of the case. This should be provided in a standardized fashion and handled the same for all internal investigations. Hanover Park utilizes a memo style form called *Internal Investigation—Report to Chief* (fig. 7–5). Attached to this memo should be the documents collected in the case file that were used to support the findings, conclusions, and any recommendations. The fire chief/CEO should be able to read this report and know exactly what happened so a decision can be made on next steps.

EMPLOYEE RIGHTS DURING THE INVESTIGATION

Employees have rights of protection when subject to internal investigations. Regardless as to whether paid or volunteer, firefighters are generally government employees (excluding contractors) who are afforded protection from the actions of the government. These protections are extended within the U.S. Constitution and Bill of Rights as detailed in the Fifth Amendment and are made applicable to state and local government through the Fourteenth Amendment. These protections are different from those extended to employees who work for private nongovernmental entities (Taylor, n.d.).

When conducting investigations, it is critical that investigators, through the delegated authority of the fire chief/CEO, understand these rights and how to proceed in order to protect both the integrity of the investigation as well as the rights of the employee. Before progressing very far into an internal investigation that has the potential to substantially impact the employee through an extended unpaid suspension or termination, investigators would do well to reach out to legal counsel for advice on how best to proceed.

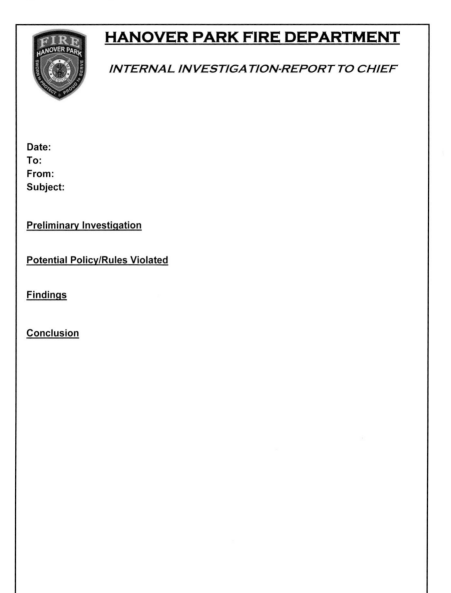

Figure 7–5. Internal investigation report to the chief (courtesy of Hanover Park Fire Department).

GARRITY RIGHTS

Garrity Rights protect public employees from being compelled to incriminate themselves during interviews conducted by their employers if they pertain to a criminal act. Garrity Rights came about based on a U.S. Supreme Court decision in *Garrity v. New Jersey* (1967). The Supreme Court held that police officers (Edward J. Garrity was a police chief in New Jersey) cannot be compelled to sacrifice their right to self-incrimination under the threat of serious discipline and to make statements that may be used in a subsequent criminal proceeding. The court later ruled in *Gardner v. Broderick* (1968) that an officer cannot be terminated solely for refusing to waive his Fifth Amendment right to remain silent. This right is also extended to firefighters.

The important distinction here is that employers have the right to question an employee about job-related issues, and an employee refusing to answer a legitimate question can be disciplined for insubordination. This also applies to sworn statements. Garrity applies when the questions migrate into the area of criminal activity. Chief Officer Curt Varone, who is both a fire officer and an attorney, explained it as follows:

> Garrity applies only when public employees are asked legitimate job-related questions by superiors and when the answers may incriminate them in a criminal matter. A firefighter in a Garrity situation has the right to remain silent, but only to the extent he or she is not ordered or compelled to answer questions. If compelled to answer legitimate job-related questions, the firefighter must do so. Any information resulting from the compelled questioning may not be used in a criminal prosecution against the firefighter. This limitation on the use of compelled information in a criminal case is often referred to as immunity. (Varone, 2019)

Chief Varone wrote in the same article,

> Contrary to popular belief, an employer has a legitimate right to ask an employee questions related to his or her work, and an employee who refuses to answer legitimate questions posed by his or her employer may be disciplined for insubordination. Answering job-related questions is not optional, and insubordination for refusing to answer legitimate questions about a job-related matter may result in termination. (Varone, 2019)

Therefore, if it is felt that the situation under investigation involves a criminal offense, then obtaining a sworn statement or conducting an interrogation as part of an internal investigation should likely be deferred until legal advice can be sought. If no criminal activity is suspected, then questioning and obtaining statements is in most cases acceptable. Again, when in question, before acting always check with labor counsel for advice to make sure that you are proceeding appropriately.

WEINGARTEN RIGHTS

For departments whose employees are represented by a collective bargaining labor organization, Weingarten Rights may be afforded. These rights are based on the U.S. Supreme Court case of *National Labor Relations Board (NLRB) v. J. Weingarten, Inc.* The ruling indicates that unionized employees have the right to request union representation at investigatory interviews where they "reasonably believe" that information learned may lead to discipline or termination (*NLRB v. J. Weingarten, Inc.*, 1975). Although the decision applies only to private sector employers, many states, as well as the federal government, have extended this right to government employees (Taylor, n.d.). Denial of an employee's request for union representation will likely draw an unfair labor practice charge against the employer. Many labor organizations go as far as to distribute cards to their membership with Weingarten Rights statements printed for use during interviews. Although the language may vary slightly, the generalized statement reads as follows:

> If this discussion could in any way lead to my being disciplined or terminated, or affect my personal working conditions, I respectfully request that my union representative or steward be present at the meeting. Without representation, I choose not to answer any questions.

When an employee exercises this right, you as the employer have a couple of options:

1. Delay the interview and wait for union representation to arrive.
   a. Provide time for the employee and the union representative to meet privately before resuming questions.
   b. Allow the union representative to counsel the member during the questioning.
   c. Allow the union representative to speak during the questioning.
2. Deny the request and end the interview.

If you deny the Weingarten request it is likely that the labor board will ultimately rescind any disciplinary action imposed by the employer. Also, any information obtained during this meeting will be excluded from any future legal action related to discipline or discharge.

A practice that I have found that works well is for the employer to actually reach out to the union and request that they attend all meetings in which an employee is going to be questioned about a topic that, depending on the answers, may lead to serious discipline (suspension or discharge). This practice by the employer eliminates the stress on the employee of having to invoke their Weingarten Rights. It also sends a message that we as a group (labor and management) need to work together to address problems and that we as the employer have no intention of circumventing the rights of employees. It is these types of proactive steps that have helped the Hanover Park department maintain strong relations between labor and management.

DISCIPLINE

> For the moment, all discipline seems painful rather than pleasant, but later it yields the peaceful fruit of righteousness to those who have been trained by it. (Hebrews 12:11)

The overarching goal of discipline is to change behavior and improve performance. Every organization has reasonable rules of conduct, whether written or implied. Rules are necessary standards used to maintain order and to allow people to work together to accomplish the mission of the organization. When a rule is violated, discipline is the tool used to modify future performance so that it aligns with the standards and values of the organization.

It is also important to recognize that discipline is hard for both the one receiving it as well as the one responsible for administering it. In all my years as a supervisor, I have lost more sleep and worried more about employee discipline than any other issue. I have also come to recognize that there often is a difference between what I personally want to do with an employee and the responsibility of the office of fire chief.

As a general tendency, I usually want to extend grace to employees who have made a mistake. I know this about myself, and I therefore must consciously remind myself of the responsibility I hold as the leader of the organization. Because of this, I often do not get to do what I want, but rather, I must do the work of the fire chief. This responsibility as fire chief sometimes involves strong discipline of lengthy suspensions and terminations. It is in disciplinary situations when the weight of the fifth bugle often feels the heaviest.

As a supervisor I think it important to understand that mistakes are made but also to recognize that for the health of the organization and the employee, discipline is sometimes required. Likewise, it is also important to recognize that discipline, administered too harshly and on a too frequent basis, where employees routinely expect discipline for even small infractions, tends to lose its effectiveness. This can then lead to employee job dissatisfaction, low morale, and high turnover. Employees become unwilling to make decisions for fear of being disciplined. They simply try to endure the job without getting into trouble. In a public safety position, this fear can lead to disastrous consequences and can have a negative impact on how we serve our communities. Our job as emergency responders is to make decisions. When our employees become unwilling to do that out of fear, we have an organizational problem. To say the least, employee discipline is a balancing act.

I also strongly believe that no employee gets up in the morning and as they get ready to come to work says, "I think I will go in and screw something up today." Mistakes happen and sometimes we have lapses in judgment, but I do not generally think that most employees intentionally do things that are malicious. I will admit though, I have seen cases where this has occurred. I just choose to think better of people until they prove to me that I should see them in a different light. We all seem to have at least one bad apple in our basket. The challenge becomes minimizing their impact.

When assessing discipline situations, the first question I always ask is what role did we organizationally play in what happened? Did we create a situation where the employee, based on our rules, regulations, standards, or techniques, faced a scenario where they were likely to make an error? Did our leaders fail to supervise? Did our training fail to prepare them appropriately? Did we as leadership model the right things? Did the employee know that what they were doing was wrong and unacceptable to the organization? When considering whether to take an adverse employment action against an employee, I think it important to first take an introspective view.

Next, I try and look at the employee's motivation. Was this a mistake or a malicious act? Was it a lapse in judgment, or is the behavior reflective of them as an individual? I consider their history and past performance and look to see if I can find any trends. Maybe they do not have any documented discipline in their past, but what have their supervisors written in their employee evaluations related to performance improvement? Also, I think about what I suspect they have done during their time working for the department. Problem employees do not suddenly become problem employees. They have a track record, and you can almost always be assured that they have done things that you do not know about or things that you might suspect but cannot prove.

As an example, maybe you have an employee who you believe is a thief. Maybe the other members of your department also believe he is a thief. Over the years whenever something goes missing, he seems to have had a level of involvement, but he has not slipped up enough that you have been able to catch him and make anything stick. If he is stealing from his employer, you can almost guarantee that he is stealing from other people as well. This is who he is, and it is a character problem that will plague your organization until he either leaves, retires, or gets caught. When thinking through employee discipline and assessing employee motivation, this type of information becomes relevant. The question is how it can be used to address the situation at hand. Or can it?

DISCIPLINARY POLICIES

Organizations need to have a policy on discipline that states how discipline is to be handled. The policy needs to spell out not the exact consequences of various infractions but rather who is covered by the policy and what standards will guide the discipline. Using Hanover Park's discipline policy as an example, the purpose section reads as follows:

> All employees, regardless of full, part-time or volunteer status, are required to exhibit acceptable conduct [see policy on Code of Conduct], satisfactory job performance and to comply with the Village of Hanover Park Personnel Rules & Regulations as well as all additional Rules, Regulations, Policies and Procedures of the Hanover Park Fire Department. Employees may be disciplined in accordance with Village policies and where applicable, said discipline will comply with the current collective bargaining contracts between the Village of Hanover Park, Hanover Park Professional Firefighters Association—Local #3452 and SEIU—Local 73.
>
> Where the interests of the Village and employee are served by progressive discipline, the Department will consider discipline that is designed to correct employee behavior. However, certain offenses are of sufficient seriousness in and of themselves that employees may be subject to more stringent action up to and including termination without any further prior disciplinary action.
>
> Final departmental disciplinary authority and responsibility rests with the Fire Chief subject to applicable grievance proceedings as stipulated by a Collective Bargaining Agreement. (Village of Hanover Park Fire Department, 2017)

Standards of performance and employee expectations are governed by the rules, regulations, policies, and procedures enacted by the department/district. These are the guiding documents that provide direction and measure performance. The discipline policy simply details how violations of these rules/policies will be handled. Take as an example a policy on patient care documentation:

> Patient Care Reports (PCRs) shall be completed and filed with the receiving hospital immediately after arrival and prior to departing the hospital. In rare occurrences/extenuating circumstances where the ambulance is required to leave the hospital prior to having the PCR complete, personnel must complete the documentation and either hand deliver or fax it to the receiving facility within 2-hours of the completion of the run. (Amboy Fire Protection District, 2020)

If an EMS provider fails to follow this policy, they have violated the district's standard on patient care documentation. After investigating the reasons behind this violation, the employer may decide to discipline the employee. The discipline policy would then be referenced to govern how this discipline is to be carried out.

CARRYING OUT DISCIPLINE

Discipline is a serious issue. It has the potential to negatively impact an employee's career, take money out of their pocket, and in some cases take away their livelihood. It also will send an organizational message. The facts of the case, how it was investigated and handled, and whether the rank and file agree with the disciplinary decision will all have an impact on the internal environment of the organization. Accounting for all these factors, an officer/supervisor should never take discipline lightly. In fact, whenever possible, officers/supervisors should work to train, retrain, and coach employees before moving to formalized discipline. This however cannot always be done based on the seriousness of the violation.

Discipline should also generally be progressive. For most infractions and for most employees, a less aggressive action will fix most situations. Progressive discipline is designed to become more aggressive should the employee continue to not perform in an acceptable fashion.

A misconception about progressive discipline is that it applies only to a repeating of the same infraction. This is not the case, and employers should not manage discipline this way. Discipline should be cumulative based on the employee's performance as a whole.

As an example, say an employee has been coached by their lieutenant a few times recently about being late for shift and not ready to work at shift change. Then the department receives a complaint about this employee being rude to a patient on an ambulance call. This reported rudeness in most cases may normally only merit a coaching session, but based on the employee's recent track record, the supervisor may elect to jump to an official documented verbal warning or maybe even a written warning.

If the employee continues to have performance problems, the discipline should progress further. The intent of the discipline is to get the employee's attention and correct the problem. Along with this discipline, the lieutenant would do well to try and figure out if there is an underlying cause for this employee's recent behavior, but in the end, the underlying reason should not negate the discipline.

Sometimes the actions of an employee are at such a level that the only reasonable intervention is to bypass lesser levels of discipline and to take a more aggressive stance (e.g., bypass a written warning and move directly to an unpaid suspension). Disciplinary action needs to fit the infraction. When writing your disciplinary policy, it is important to state in the document that the department/district reserves the right to give a more stringent discipline based on the seriousness of the situation. This will make the policy clear to the employee and will give the department/district the latitude needed to effectively govern the organization.

Using my earlier example of the employee arrested for driving under the influence, a coaching session or a documented verbal warning would not align with the severity of the infraction. A much more severe action is required. However, the employee showing up late and not being ready for shift probably should not result in a two-week unpaid suspension. The level of discipline needs to fit the crime.

PROGRESSIVE DISCIPLINE

When writing your organization's discipline policy, I recommend that you spell out each step of the disciplinary process, including who within the supervisory ranks has the authority and responsibility to take disciplinary action.

This issue of responsibility, in my opinion, is an area of concern for the fire service. It is not a new concern but one where I have watched officers repeatedly fail throughout my career, both volunteer and paid. We want to be officers, but we do not want to discipline our friends. We understand that discipline is designed to correct behavior and to steer the employee in the right direction; we just want someone else to do it. If this is your practice, let me be clear: you are not fulfilling your duties as an officer/supervisor.

Likewise, for fire chiefs/CEOs, if you are not delegating discipline to your lower-level supervisors, you also are not fulfilling your duties appropriately. Employee performance problems should be solved at the lowest organizational level possible. When I served as fire chief, I would tell my people, if a disciplinary situation is of such significance that it gets to my level, somebody is probably going to get fired. All other discipline should be handled at a level well below the office of fire chief (Haigh, 2019).

Using Hanover Park's discipline policy as an example, here is how we outlined our process.

- **Verbal Warning** (Administered by any departmental supervisor or acting supervisor)
  - *Employee Coaching*—A supervisor has the ability and authority to "Coach" an employee in hopes of correcting specific behavior or performance. The supervisor should make a note in the employee's evaluation log as a reminder of the coaching and employee expectations or outcomes.
  - *Documented Verbal Warning*—When a supervisor deems it necessary to correct a specific behavior or performance and feels that the infraction is of a serious enough nature to warrant a documented verbal warning, such shall be completed and subsequently documented on the Village of Hanover Park's Disciplinary Action Form and submitted to the employee's Battalion Chief and then to the Assistant Chief and Fire Chief. The consequence of not correcting the behavior or performance is more progressive disciplinary action. The employee may refuse to sign the Disciplinary Action Form. Their refusal to sign does not negate the discipline.
- **Written Warning** (Administered by any departmental supervisor or acting supervisor)
  - When a supervisor deems it necessary to correct a specific behavior or performance and feels that the infraction is of a serious enough nature to warrant a documented written warning or is a repeated violation, such warning shall be completed and subsequently documented on the Village of Hanover Park's Disciplinary Action Form and submitted to the employee's Battalion Chief and then to the Assistant Chief and Fire Chief. The consequence of not correcting the behavior or performance is more progressive disciplinary action. The employee may refuse to sign the Disciplinary Action Form. Their refusal to sign does not negate the discipline.
- **Suspensions** (Administered by the Assistant Fire Chief or Fire Chief)
  - A Chief Officer can/shall recommend suspension(s) to the Fire Chief or Assistant Chief for behavior or performance of sufficient

seriousness or repetition. In the absence of the Fire Chief and Assistant Chief, a Chief Officer can/shall administer the suspension(s) or administrative leave(s).
- With Suspensions lasting longer than seventy-two hours, the employee will be required to surrender their badge and Village of Hanover Park Identification.

- **Demotions and Terminations** (Administered by Fire Chief)
  - Demotions or terminations may be administered in response to behavior or performance when lesser forms of discipline have not resulted in improved behavior. The Fire Chief reserves the right to demote or terminate an employee due to severe infractions. Demotion to a lesser or lower rank will also result in a corresponding reduction in wages or salary. The demotion process does not preclude termination or other forms of discipline; it only provides an additional disciplinary option.
- **Relief of Duty** (Administered by a Chief Officer or acting Chief Officer i.e. lieutenant acting as a battalion chief with immediate notification to the Fire Chief)
  - Pending further investigation or disciplinary action, a chief officer may relieve an employee of duty and order them to leave the premises if his/her continued presence would:
    1. Disrupt the inner workings of the department
    2. Undermine the public's trust in the department
    3. Affect the health and safety of the employee, co-workers, or the public
  - Upon ordering the employee from the premises, the Chief Officer shall immediately notify the Fire Chief. The supervisor shall document the circumstances of the interim relief of duty using the Village of Hanover Park's Disciplinary Action Form for submission to the Fire Chief. The Fire Chief and charging Officer shall meet with the employee within a reasonable amount of time to investigate and determine the employee's status.
  - An employee may be placed in an administrative leave status, with or without pay, any other time an employee must be removed from the workplace until a proper investigation or other administrative proceeding is completed. During periods of Administrative Leaves lasting longer than seventy-two hours, the employee will be required to surrender their badge and Village of Hanover Park credentials. During periods of administrative leave, the employee shall continue to comply with all policies and lawful orders of a supervisor. (Village of Hanover Park Fire Department, 2017)

DOCUMENTATION

Documentation is an essential step in the disciplinary process. The creation of a paper trail is necessary to support further disciplinary action. This documentation needs to be available to other officers/supervisors so they can effectively manage respective employees. The Hanover Park Fire Department utilizes a standardized *Disciplinary Action Form* for documenting discipline (fig. 7–6).

For documenting employee coaching/counseling, instead of completing the disciplinary action form, a note is made in the employee journal section of the employee performance evaluation software. Departments/districts not using a formalized software system can accomplish the same thing by simply writing a "note to file" and placing it in the employee's personnel file.

Something that I have used in the last few years for my direct reports is a follow-up email after a coaching conversation detailing what we talked about and the agreed upon corrective action. I then place a copy of this email in the employee's personnel file. The takeaway message—do not get hung up on the process of documentation, but rather, do the documentation. Even if your organization does not have a formalized or set standard related to this issue, you as the officer/supervisor need to find a system that works for you.

TERMINATIONS

As much as our goal should be to not get to a position where we need to terminate an employee, sometimes it happens. Employees make mistakes, and sometimes these mistakes lead to their discharge from the organization. Termination cases are typically fraught with anxiety, stress, and the full knowledge that legal action is likely to result from the dismissal. These legal proceedings will review every decision and step under a microscope in an all-out battle to try to paint a picture that you, as the boss, made the wrong decision. The attacks may be vicious, with accusations that you are incompetent, you had it out for the employee, and you heartlessly destroyed the calling of a true public servant. Sound fun? It is not! Part of the job? Unfortunately, yes!

Bottom line, it is hard to fire someone—and it probably should be.

CASE FILES

When it comes to the actual decision of whether to move forward with a termination, this action ought not be made without wise counsel and in many cases should be a collective decision made by senior department/district leadership, including human resources, labor attorneys, and city/village/district/township managers/supervisors. To facilitate this decision, in my community

Village of Hanover Park Disciplinary Action Form – Fire Dept.

Employee Name		Position		Date of Warning	
Employee/Payroll #		Department		Division	

Type of Violation

☐	Absenteeism	☐	Carelessness	☐	Insubordination
☐	Lateness or Early Quit	☐	Failure to Follow Instructions	☐	Violation of Safety Rules
☐	Rudeness to Employees or Customers	☐	Willful Damage to Material or Equipment	☐	Working on Personal Matters
☐	Unsatisfactory Work Performance	☐	Violation of Village or Department Policies or Procedures	☐	Other

Employer Statement (attach all supporting documentation)

Date of Incident ___/___/___ Time_____

Employee Statement

☐ I agree with Employer's statement.
☐ I disagree with Employer's description of violation for these reasons:
_____
_____
_____
_____

Action to be taken   ☐ Oral Warning   ☐ Written Warning   ☐ Probation   ☐ Dismissal
☐ Suspension: ___ day(s) from ___ through ___   ☐ Other_____

Consequence should incident occur again

I have read and received a copy of this Disciplinary Action form and understand it.

_____   ___/___/___   Copies to:_____
Signature of Employee

_____   ___/___/___   _____
Signature of Supervisor Who Issued Warning

_____   ___/___/___   _____
Signature of Department Head

Immediate satisfactory improvement must be shown or further disciplinary action will be taken, including possible suspension from duty or discharge

ROUTING: Original to Human Resource Department (Employee Personnel File), Copy to Employee, Copy to Supervisor

Figure 7–6. Disciplinary action form (courtesy of Hanover Park Fire Department).

I am required to build a case file detailing all information related to the situation, including the internal investigation, interrogations, past discipline, coaching, employee training, and the like. The goal in compiling this file is to justify through supporting documents the proof needed to support a decision for dismissal. In my situation, this case file, once complete, is presented to the village manager and labor attorneys for review and oversight.

When creating these case files, I tend to write them as if I am writing a research paper. I begin with an abstract that summarizes the major aspects of the case. I provide a general summation of the investigation, the actions taken, and my recommendation for termination. I try to keep this section to between 300 and 500 words designed to give the reader a high-level overview of the situation.

In the next section I delve into the details of the case. I usually cover this section using a timeline format. Within the timeline I reference and then attach documents to support my statements (e.g., reports from law enforcement, chemical analysis reports, sworn statements, photos).

In cases where I am trying to justify termination for poor job performance, I often use the employee's job description as the basis. In these cases, I walk through the requirement/expectations as outlined in the job description, making detailed comments as to how the employee has failed to perform at an acceptable level. With each one of these comments, I reference and attach documents showing where the employee has been trained, coached, retrained, directed, and disciplined. My goal is to show that we as the employer have done everything possible to help the employee perform to an acceptable standard.

Creating these case files takes many hours. They are frequently, when all attachments are included, very large documents. Once this report is complete, I would send it to our legal counsel and my assistant chief/executive officer and ask them to read the document with a critical eye to find any holes in the case. Sometimes I would also send it to one of my trusted confidants for an outsider's view. These reviews help to provide a perspective on whether I have truly made the case for termination. Often these reviews come back with questions and sometimes an analysis that says I do not have enough evidence to support the recommended discipline. That is exactly how these case reviews are supposed to work. Your team is supposed to try to punch holes in the case. It is far better to find weak areas before you make an adverse employment decision than to face these same questions as a defendant on the witness stand. I also want to know from my legal counsel whether they can defend and win the case should it go to arbitration or court.

LOUDERMILL HEARING

Once it has been decided that termination is the correct move, public sector employees are afforded the right to a predisciplinary hearing where the employer is required, normally in writing, to provide notice of the charges against them, as well as an explanation of the evidence, and then allow an opportunity for the employee to present their side of the story. Loudermill hearings also apply in cases where the employee will be suspended without pay as well as in demotion cases (Corbin & May, 2012).

The background of Loudermill hearings comes from the U.S. Supreme Court case *Cleveland Board of Education v. Loudermill.* In this case the board of education hired Loudermill as a security guard. The security guard job was designated as a "classified civil service" position. Loudermill, in his job application, stated that he had never been convicted of a felony. However, it was later discovered that he had in fact been convicted of grand larceny. The board of education fired him for dishonesty. Due to his civil servant status under Ohio law, Loudermill could only be terminated for cause. The court held that due to his civil servant status he was entitled to a hearing prior to termination to explain and refute any conclusions being used to justify the decision for termination (*Cleveland Board of Education v. Loudermill*, 1985).

To maintain compliance with the requirements of Loudermill, do the following:

- Notify the employee well in advance of the meeting so they have sufficient time to have a representative present.
- Provide a written notice of the meeting, including the charges, evidence, and the discipline being contemplated. Tell the employee that this is their opportunity to provide information as to why the proposed discipline should not occur.
- Allow the employee to speak and to provide their perspective on the situation. Often employees bring a written statement that they wish to read during the hearing. Sometimes their union representative or legal counsel will speak on their behalf. As the employer, it is fine to ask clarifying questions, but do not cross-examine the employee.
- Do not engage in behavior that indicates agreement or disagreement with the employee's position. You are just here to listen.
- Do not feel compelled to allow witnesses (unless this is a requirement within the employee's collective bargaining agreement).
- If new information is presented by the employee that needs follow-up, this should be done prior to a final decision.
- There is no obligation to make a final decision on the day of the hearing. Consider what you have learned and weigh it against the facts. Then proceed with a timely written decision following the hearing (American Federation of Government Employees Local 704, n.d.; Corbin & May, 2017).

THE REST OF THE STORY

I opened this chapter with a story. The story speaks of an employee's lapse in judgment and her subsequent arrest. It also is a great example of the weight of the fifth bugle and the sometimes-awful decisions we are required to make. The story is not fictitious and certainly does not apply to the Hooterville Fire Department but rather the Hanover Park Fire Department. It also provides a couple of key learning points for employee discipline:

- First, it exhibits the importance of relationships and how the initial notification of the situation came to me through the president of the full-time firefighters' union.
- Second, it shows how the chain of command should work in these situations as the information was passed from the on-duty battalion chief through the chain and ultimately to the mayor.
- Third, it speaks to the need to get in front of these types of issues before they grab the attention of the media.
- Fourth, it talks of the stress and the intense weight I personally felt based on the fact that this employee is indeed one of my favorites (I know that fire chiefs are not supposed to have favorites—I guess that I fail in that category).

So here is the rest of the story...

Based on the information I learned, I immediately ordered the initiation of an internal investigation and reached out to our labor counsel for assistance. The team interviewed the employee, and because of our department's culture of telling the truth, the employee through tears provided details of every aspect of the situation. As the investigation continued to unfold, at no point did we uncover a single discrepancy in the information the employee provided. Based on the alleged misconduct, I made the decision early to "stop the clock" and place the employee on paid administrative leave and to order that she surrender her credentials.

The investigating team requested copies of all reports and videos from the arresting agency and interviewed the arresting officers. The employee was formally interrogated with both a union representative and an attorney present, following all requirements of the Illinois Fireman's Disciplinary Act (1983). A court reporter took down a transcript of what was said. Based on the findings, I compiled a case file and submitted a recommendation to the village manager charging the employee with "actions unbecoming" and recommended termination. The village manager agreed, and a Loudermill hearing was held. The employee was terminated.

As expected in these types of cases the union filed a grievance. The grievance quickly progressed through the step process as outlined by the collective bargaining agreement (i.e., Step 1: Request for reconsideration by the fire chief; Step 2: Request for reconsideration by the village manager) and then on to arbitration. The case was heard by an arbitrator almost 10 months later. After a full day of testimony by several witnesses, the arbitrator ruled that I as fire chief had overstepped, that the termination was too punitive, and that the employee was to be reinstated. He also ruled, however, that due to the severity of the misconduct, although it did not rise to the level of termination, significant discipline was warranted. Based on the latter, he held that the employee's time off (almost 10 months since the termination until the arbitration hearing) was to be considered an unpaid suspension. Along with the suspension was a loss of seniority for the same amount of time. As I had predicted from the start, there would be no winners here . . . only losers.

The employee returned to work, which was incredibly awkward for all involved. I met with the employee as well as the union and detailed performance expectations going forward. I stressed that the employee did not need to be fearful of future retaliation in that we as an organization would support the decision of the arbitrator.

Also, because of the nature of the offense, there were fire department employees on both sides of this situation. Some supported the arbitrator's decision to return her to duty, while others were angry that their union dues had been used to defend an employee who clearly violated the law and had given the department a black eye with a neighboring law enforcement agency. The union president and I talked about these feelings openly and the challenges we both faced. Due to his strong leadership, he was quickly able to minimize this situation's impact and to get the team again focused on the mission of the department.

Another critical aspect that helped in moving forward was how the returning employee handled the situation. Because of her character and through deep reflection, the overall situation resulted in a life-altering course correction. This individual was a good employee before this event, but I would argue that today she is an exceptional employee. This employee now serves as the department's coordinator of public education training, responsible for management of CPR, First Aid, and Stop the Bleed classes. She developed a program where Stop the Bleed kits were donated to local schools and teachers were trained to provide life-saving measures during an active shooter event. The employee now also serves as a field training officer and openly shares her story with new department members. Having the courage to openly share this story will likely prevent similar situations with other employees for years to come.

My friend Curt Varone, during one of our conversations related to disciplinary action and fire service legal issues, commented that this case is an example of how discipline is supposed to work—it changed behavior and improved employee performance.

Although painful for all involved, I think that Chief Varone's assessment is correct!

REFERENCES

Amboy Fire Protection District. (2020, February 1). *Administrative Policy: Incident Report Documentation*—1.1.31. Amboy, IL.

American Federation of Government Employees Local 704. (n.d.). *Loudermill Rights*. https://afgelocal704.org/your-rights/loudermill-rights/.

Cleveland Board of Education v. Loudermill, 470 U.S. 532 (1985). https://supreme.justia.com/cases/federal/us/470/532/.

Corbin, J., & May, J. (2012, March 1). *Taking the Mystery Out of Loudermill Meetings*. MRSC. http://mrsc.org/Home/Stay-Informed/MRSC-Insight/Archives/Taking-the-Mystery-out-of-Loudermill-Meetings.aspx.

Fireman's Disciplinary Act, Ill. Stat. § 50 ILCS 745 (1983). http://www.ilga.gov/legislation/ilcs/ilcs3.asp?ActID=740&ChapterID=11.

Gardner v. Broderick, 392 U.S. 273 (1968). https://supreme.justia.com/cases/federal/us/392/273/.

Garrity v. New Jersey, 385 U.S. 493 (1967). https://supreme.justia.com/cases/federal/us/385/493/.

Haigh, C. A. (2019, July). Challenges Behind the Officer's Collar Brass. *Fire Engineering*. https://www.fireengineering.com/leadership/challenges-behind-the-officer-s-collar-brass/.

Hanover Park Fire Department. (2017, April 1). *Discipline Policy: SOG—100-006 Discipline*. Hanover Park, IL: Village of Hanover Park.

National Labor Relations Board (NLRB) v. J. Weingarten, Inc., 420 U.S. 251 (1975). https://supreme.justia.com/cases/federal/us/420/251/.

Taylor, D. (n.d.). *Garrity Basics*. Garrity Rights. http://www.garrityrights.org/basics.html.

Varone, J. C. (2019, January 1). Fire Law: Garrity Update: A Firefighter's Right to Remain Silent. *Firehouse*. https://www.firehouse.com/careers-education/article/21032362/fire-law-garrity-update-a-firefighters-right-to-remain-silent.

Varone, J. C. (2014). *Legal Considerations for Fire & Emergency Services*, 3rd ed. Tulsa, OK: Pennwell Corporation.

QUESTIONS FOR FURTHER RESEARCH, THOUGHT, AND DISCUSSION

1. What is your organization's policy related to receiving and processing complaints?

    a. Where do you see room for potential legal challenges?

    b. What recommendations would you make related to changes or improvements in the complaint process?

2. By policy, who in your organization has the authority to direct an employee to write a sworn statement?

3. Who in your organization is tasked with the responsibility to conduct or manage internal investigations?

    a. What range of authority do they have related to conducting employee interrogations?

b. In what type of cases, either by policy or practice, does the investigating officer have the authority or expectation to bring in an outside expert to assist with case management (law enforcement, legal counsel, contractor investigator)?

_____
_____
_____
_____
_____

4. Based on the statutes/regulations/laws of your state, what rights are extended to employees of your organization based on the case *NLRB v. J. Weingarten Inc.* (1975)? What is your process for extending Weingarten Rights if applicable?

_____
_____
_____
_____
_____

5. Is your organization governed by a state or federal regulation related to firefighter discipline (firefighter disciplinary act)? If so, provide a paraphrased overview of the requirements, including an analysis of its applicability to your specific organization.

_____
_____
_____
_____
_____

6. Per your organization's policy, who has the responsibility and authority to administer discipline at each of the following levels:

   a. Employee coaching: _____
   _____

   b. Documented verbal warning: _____
   _____

   c. Written warning: _____
   _____

   d. Suspension: _____
   _____

e. Demotion: _____

f. Termination: _____

g. Relief of duty: _____

h. Administrative leave (paid or unpaid): _____

7. Per your organization's policies, including any applicable collective bargaining agreements, what rights are extended to employees to request a review and reconsideration of discipline?
_____
_____
_____
_____

   a. If a formalized grievance process is allowed, what are various steps and who is involved in the decision-making process?
   _____
   _____
   _____
   _____

   b. Who has the final decision-making authority related to reviews of employee discipline issues?
   _____
   _____
   _____
   _____

8. How does your organization document and track employee discipline?
_____
_____
_____
_____

9. Review your organization's policies, including applicable local ordinances, state statues, federal regulations, and collective bargaining agreements, and then detail the steps required to process and hold an employee Loudermill hearing, including any required employer response and associated timelines.

_____
_____
_____
_____
_____

CHAPTER 8

POLICY DEVELOPMENT

In the spring of 1989, I had just been promoted to lieutenant at my volunteer fire department. The fire chief at the time, William Kruse, mandated that all his officers attend a presentation given by Chief Alan Brunacini of the Phoenix (AZ) Fire Department. Chief Brunacini was teaching fireground management based on his book *Fire Command* (fig. 8–1). He talked about the need to create standard operating procedures or a "playbook" to provide guidance on how we manage incidents. Although I had heard the phrase "standard operating procedures," it was a relatively foreign concept as neither of my departments (paid or volunteer) had written procedures focused on the areas he was describing. I had always been taught that incidents are dynamic and vastly different, and the creation of a preincident playbook was pretty much nonsense. Chief Brunacini argued just the opposite. In an interview with Robert W. Grant, conducted on behalf of the U.S. Fire Administration, Chief Brunacini talked about the use of standard operating procedures within the Phoenix Fire Department:

> The [standard operating] procedures provide an organizational plan. In other words, it's a set of organizational directives that establishes a standard course of action on the fire ground, and it describes the way we command, the way we [sectorize], the standard company functions of every different kind of unit, the safety procedures that we use, and that's a high priority that we have, the way we communicate with one another, and the basic organization that we're going to assume. And they're written very specifically for our own organization. Although, I think that there's a lot of common elements in them. We started that development six or seven years ago now. I think it has just really had a dramatic impact on the effectiveness of our ability to go out and extend those services in the field. (Grant, n.d.)

I was intrigued to say the least. In fact, to use the word "impressed" with Chief Brunacini and his commonsense approach to fireground operations may be a vast understatement. As I left this training, I continued to think about written policies. I really had no idea what they looked like, how they were constructed, what information or directives they contained, or exactly how to implement them. I just knew that Chief Brunacini thought that they were important, so I assumed that I should think they were important.

Figure 8–1. The author's well-used and autographed copy of Chief Brunacini's book *Fire Command.*

I remember talking to my fellow officers following the training. We all agreed that we needed something more than what we currently had. I had joined the volunteer department in 1983, and we had basically been operating from a one-page document called "standing orders." This document focused primarily on what apparatus to take to certain types of calls and how to provide a fireground size-up:

- Code Green—Nothing showing
- Code Yellow—Smoke showing
- Code Red—Flames showing
- Code Black—Your apparatus was disabled and was no longer serviceable (We did not want the folks in radioland to know that our rig had just died.)

I felt that if I could see what others were using and what standard operating procedures actually looked like, I could probably create something for us to work from. I began searching for examples, and through a variety of fire service contacts I was able to get documents from FDNY, Boston, Los Angeles County, and Phoenix. Using these documents as a template I crafted Hampton's first set of standard operating procedures (now referred to as guidelines or SOGs). We then used these documents as the basis for our weekly training sessions. It took some time, but this "written thing" caught on, and our guys started relying on the documents to assist in their decision-making.

Today it seems strange to think that departments would not have some level of written policies. But back then, at least in my area, this was the norm. Chief Brunacini's teaching that day changed my thinking and subsequently how my department operated. The change made us more efficient and safer and certainly enhanced our quality of service to the community. I credit this change to Chief Brunacini, now deceased. I am ever so thankful that as the years moved forward, he and I became friends.

GUIDING DOCUMENT

All emergency service organizations have two things in common: they have employees, and their employees respond to calls. Based on this, I believe that three guiding documents are an absolute must:

- Employee handbook
- Standard operating guidelines (SOGs)
- Job performance requirements (JPRs)

The employee handbook is a compilation of policies intended to provide direction for both the organization and the employee. It should clearly define the expectations of performance, including the rules of conduct. Handbook policies are designed to be inwardly focused.

SOGs on the other hand are outwardly focused. They are designed to give direction on how the organization provides services. SOGs operate from the premise that emergency incidents are more alike than different and consistency in performance reduces mistakes.

Some organizations use the word "procedures" instead of "guidelines" to describe these documents (standard operating procedures versus standard operating guidelines). Either is fine since they both describe how services are to be provided. Some will argue that the word "procedures" is legalistic and does not provide any sense of leniency within an operation. They hold to the position that any variance would increase legal liability; therefore, they believe the change to "guidelines" better reflects the dynamic nature of decision-making. I have talked to attorneys who suggest that this word change makes a difference, while others say either phase is fine. I usually call them guidelines, but I think this argument is best left with the individual agencies themselves.

We often refer to these compiled policies as books or manuals (i.e., employee handbook or SOG manual). I think this comes from our historical practice of printing the documents and then compiling them in a three-ringed binder. This is the practice most of us have followed for years, and once you get all the documents together, the binder indeed looks like a book. Although this system works, I think that today it is better to maintain the files electronically with searchable options that allow employees to quickly access the documents via their smartphone or tablet.

Both books and manuals are designed to provide direction and to be a source of reference while preventing problems through clear communication. The policies say, "This is how we do it here." When problems occur or our personnel do not perform at their best, we need to ask ourselves the following questions:

- Do we have a policy addressing the topic?
- If so, what does it say?
- Did our personnel know about the policy?
- Did they receive training on the policy?
- Can we prove that they knew about the policy and had received training (i.e., do we have an employee-signed acknowledgment form)?

Another series of documents that complement SOGs is JPRs. SOGs focus on operations while JPRs focus on skills. Rock Island (IL) Fire Department calls their JPR book the *Skills and Maintenance Manual*. I actually like this name better. I think it is a simplified and more accurate reflection of how these documents are to be used. Think of it like this:

- Job performance requirements (JPRs) are used to teach the skills needed to carry out operations as identified in the SOGs.
- Standard operating guidelines (SOGs) provide guidance on when to implement the skills taught through the JPRs.

JPRs will without question make an organization more efficient by providing a standardized training plan for teaching and evaluating the skills we use on the job. If I were starting from scratch and working to develop a department's guiding documents I would likely work in this order:

- First—SOGs
- Second—JPRs (i.e., skills and maintenance)
- Third—Employee handbook

FIRST THINGS FIRST

If SOGs are new to your organization, or you have been assigned the task of doing a policy update, focus first on the topics of greatest immediate need. When I started writing Hampton's SOGs my initial work covered areas such as emergency response, use of protective equipment, engine company operations, and the basics of fireground support (i.e., ladders, overhauls, ventilation, forcible entry, search/rescue, salvage, and control of utilities). I knew that we would ultimately need to expand and refine these preliminary SOGs—but we needed to start somewhere.

STANDARD OPERATING GUIDELINES

Writing SOGs can be challenging. Many writers often find themselves with a level of "analysis paralysis" as they work to determine what information to include, the level of detail, and the overall format for the documents. Added to this is the realization that you as the writer rarely know everything about a particular topic, and the need for study and research becomes an essential, if not the most important part of the development process. Lastly, writers would do well to consider the instructional words of Robert-Louis Stevenson, the famed author of *Treasure Island* and *The Strange Case of Dr. Jekyll and Mr. Hyde:* "Do not write merely to be understood. Write so you cannot possibly be misunderstood." I personally find his advice to be profound when considering the importance of writing instructional documents for use by emergency responders.

When writing SOGs I always find it helpful to read and study the policies of other agencies. What they have chosen to include, the specifics of how they operate, and their sources of reference are all great learning tools. Although it is acceptable to borrow bits and pieces from other agencies, do not fall into the trap of copying another agency's policy and then simply replacing their name with yours. SOGs need to be individualized based on your apparatus, how you work, how your people are trained, and what makes sense for the response district you serve. I have watched many organizations take the lazy path of doing a copy and paste, change-the-name approach only to find that they have a very impressive looking SOG manual that is not reflective of how they operate. If operations do not align with the SOG manual, then the document is useless, and the work involved was a waste of time.

CATEGORIZING SOGS

When setting up SOGs I like to break the policies into broad categories with divisional headings. I think this helps in tracking and cataloging the documents, and it makes the search process easier for your personnel. In Hanover Park we divided our SOGs into several broad categories. This helps to identify the importance of the policies for the users. As an example, the operational and EMS policies are critical to know and, in some cases, memorize. The administrative, training, inspectional services, and emergency management policies require a general understanding of the documents and can be referenced when needed.

- 100—Administration
- 200—Operations
- 300—Emergency medical services
- 400—Training
- 500—Inspectional services
- 600—Emergency management

Within each of these categories, individual policies are then given an identifying number and title:

SOG 200-007 General Fireground Tactical Operations

- The SOG is part of the Operations section (200)
- The policy is number seven (007) of those included within this section
- It addresses General Fireground Tactical Operations

Another example:

SOG 100-018 Post-Incident Analysis

This policy is part of the Administrative section, is policy number eighteen, and addresses how we will conduct the process of Post-Incident Analysis.

I am not aware of any hard and fast rules related to the categorization of policies or even what categories are used. Rather, each department/district needs to assess what will work best for them. When I categorized Hanover Park's policies, I tried to place like topics together as much as possible. Table 8–1 has a few examples.

Table 8–1. Categorizing policies for each department.

100 Administration	200 Operations	300 EMS	400 Training	500 Inspectional services	600 Emergency management
Vehicle accident reporting/investigation	General fireground tactical operations	EMS system protocols	Live fire training	Annual fire safety/business license inspection	Emergency operations plan
Photo and video documentation	Engine company operations	Bloodborne pathogen/exposures	Firefighter retraining standards	Annual pre-fire survey	Incident action plans (IAPs)
Public information officer (PIO)	Truck company operations	Helicopter transports	Job performance requirements (JPRs) testing	Public education events	Incident command system (ICS) 214 logs
Post-incident analysis	Tender company operations	Mass casualty	Training documentation	Issuance of building permits	Emergency Operations Center (EOC) operations

WHAT IS INCLUDED

I find it helpful to use a standardized format (i.e., template) for SOGs. This gives them a uniform look, but more importantly it makes sure that each policy covers a minimum of information. This is incredibly important when you have several different personnel working on creating your SOGs. It is also important to have multiple reviewers read through the document to make sure that the policy accurately reflects your organization's practices. I also find it helpful to assign one individual the responsibility for the finalized formatting and publishing of the documents. This ensures consistency and eliminates the seemingly always present challenge of duplicate or missing numbers. In Hanover Park this is handled by the chief's administrative assistant.

Figure 8–2 shows the policy template we use in Hanover Park.

- Issue Date/Revision Date
  Having SOGs dated is important for a couple reasons.
    - It allows you to get the document into a rotational cycle for review and update. I like to review SOGs on a five-year cycle to ensure that they remain valid and continue to accurately reflect how we operate. The key is to have policies in a constant state of analysis and revision to ensure that we actually do what we say we do.

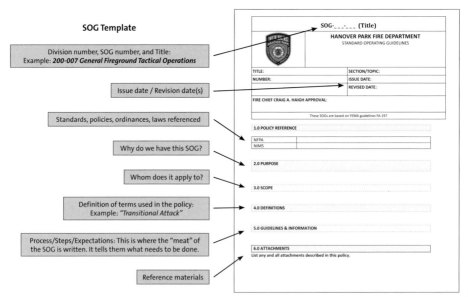

Figure 8–2. A standard operating guideline template (courtesy of Hanover Park Fire Department).

- For litigation purposes related to employee injuries or discipline, it is critical to know which edition of a policy was in place at the time of the event. Along with this, make sure that when a policy is changed or updated, copies of all prior editions of the document are maintained so they can be referenced later. This becomes important since investigations and litigation hearings and trials often occur years down the road.

- Purpose
  This section describes why the policy exists. For example, this section in our Firefighter Rehab Policy reads as follows:

  > To ensure that the physical and mental condition of members operating at the scene of an emergency, training exercise, or other department activity does not deteriorate to a level that affects the safety or well-being of each member or that jeopardizes the safety and integrity of the operation. (Hanover Park Fire Department, 2003/2007/2012/2017)

- Scope
  This section details who is covered by this policy.

  > This guideline shall apply to all activities of Hanover Park Fire Department, including, but not limited to, fire ground operations, EMS operations, training exercises and drills where strenuous mental and physical activities or exposure to heat or cold exist. (Hanover Park Fire Department, 2003/2007/2012/2017)

- Definitions
  This section defines terms or phrases that will be used throughout the policy.

  > **Company Level Rehab:** A rest and rehab period following moderate exertion that typically involves the wearing of some level of PPE. Rehab is implemented with fluids and food products carried on response apparatus and handed out at the scene. Personnel are tasked with the responsibility of monitoring each others' overall physical condition without providing formalized rehab that involves core body cooling and medical monitoring.

Formalized/Full Rehab: Establishment of a specified division of operation referenced as Rehab. This division provides an area conducive for firefighter rest and recovery, hydration, nourishment, cooling (including core body when applicable), medical monitoring and accountabilty/documentation. Each firefighter assigned to the rehab division typically spends 10–30 minutes in a physiologic recovering mode while assigned to this division. Release from the divison and return to work is based on a detailed medical exam criteria. (Hanover Park Fire Department, 2003/2007/2012/2017)

- Guidelines & Information
  This is the section where the specific details of performance are written and directives are given. This generally includes the following:
  - Who is responsible to do something
  - When is it supposed to be done
  - How is it supposed to be done (i.e., performance standard)
  - Any special information (e.g., hazards, required PPE, specialized tools)
- Attachments
  Any specialized reporting forms or reference information that would be helpful while performing the tasks associated with the policy.

EMPLOYEE HANDBOOK

Like creating SOGs, developing employee handbooks is a challenging task. In my opinion, what makes them especially hard is the fact that they are in many ways a legal contract between the employer and the employee (even though in our disclaimer statements we tell our employees not to view them as such). They define expectations for the employee, but they also spell out what the employer will do for the employee. This includes benefits, compliance with employment laws, workplace safety, and a myriad of other topics. Anytime an employee is the subject of an adverse employment decision based on a standard established within the handbook, the potential exists for legal review by either an arbitrator or judge.

Most handbooks contain a clarifying statement or disclaimer at the beginning of the book to clarify how the policies will be applied. An example of a statement may read as follows:

> This handbook is designed to provide highlights of policies, practices, and benefits for your personal understanding and cannot, therefore, be construed

as a legal document. It is intended to provide general information and is not intended to be an expressed or implied contract. Since it is not possible to anticipate every situation or question that may arise in the workplace, this document is not intended to be a substitute for sound employer management, judgment, and discretion.

The department further reserves the right, at its sole discretion to modify or delete certain aspects of this handbook as it become necessary from time to time.

Should any provision in this handbook be found to be unenforceable and/or invalid, such finding does not invalidate the entire handbook, but only the subject provision.

LEGAL VALIDITY OF POLICIES

Policies need to describe the performance standards of the organization, set expectations, and be legally sound and not discriminatory in any way. As an example, if a department maintains a policy that says male employees must maintain their hair length so that it does not touch the collar of their uniform shirt yet exempts female employees from this "hair length" standard, this is potentially problematic. Unless a basis can be established saying that hair length is a safety concern and the standard is applicable to all employees doing the same job, both male and female, this standard could be argued as discriminatory toward male employees. Another example would be a prohibition on visible tattoos. The ability of an employee to do the job in an excellent fashion has nothing to do with whether they have tattoos. It does however have applicability to the establishment of a uniform dress policy. Therefore, employers cannot discriminate against an employee with tattoos, but they may require the employee to cover the tattoos when working for the employer.

Also challenging is the fact that lawmakers are constantly establishing new laws related to employee rights. Unlike SOGs that can be reviewed on a longer cycle, employee handbooks need to be under a constant review process. As employers become aware of the establishment of a new law, they need to review how it applies to their organization and then make modifications, as necessary.

Probably the greatest difficulty in trying to stay on top of changes to employment standards is the impact of case law. Laws and employer policies are constantly being challenged in the courts, and the case law established by these trial decisions provides direction on how to implement employment standards. The task of continually researching, reading the decisions, and figuring out applicability to your organization is literally a full-time job. Since most fire chiefs/CEOs do not have the luxury of in-house counsel or the ability to continually monitor these laws and cases themselves, I highly recommend that departments/districts contract with a legal firm specializing in employment law. There are also organizations such as the National Public Employer Labor Relations Association (https://npelra.org/) that monitor and provide advice on these issues. These associations usually provide training, best practice advice, and policy templates that can be a huge source of assistance to public sector employers.

Lastly, when I write employee handbooks and or make modifications to policies, I routinely have these checked by our labor attorneys. It is extremely easy to have blind spots when creating these policies and to miss loopholes that exist related to enforcement. Having a review by someone who understands the law and will also be responsible to defend this policy in the future is beneficial and highly recommended. Bottom line, when it comes to writing and maintaining policies that apply to employment practices, it is best to not try to go it alone. This is where spending a few dollars up front will save you huge dollars down the road.

TEMPLATES

I routinely see employee handbooks written using two variations of formatting. One utilizes a running category format where topics are covered one right after another (fig. 8–3). Some organizations list an issue date on or near the front cover and every time a policy is updated, a revision date is added. This practice is somewhat misleading in that it gives the impression that the entire manual has been updated. I have seen other handbooks list revision dates next to each category as it is modified. I think this is a better practice, as shown in this example:

Use of Intoxicating Substances (Rev. Dates 6/2015, 7/2018, 1/2020)

Just as with SOGs, for the purposes of litigation (e.g., employee discipline, discrimination charges, charges of inappropriate rule applications) it is critical to know which edition of a policy was in place at the time of the event. Therefore, when making revisions, make sure that copies of the prior documents are maintained in a file so they can be referenced later if needed.

Figure 8–3 is an example of a handbook written using the running category format.

Chapter 8 Policy Development

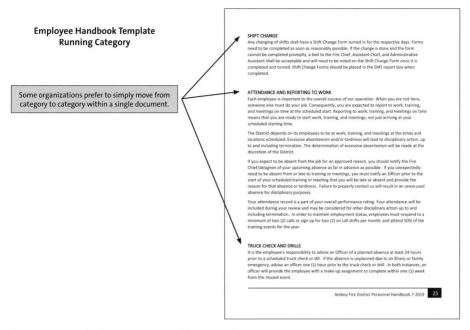

Figure 8–3. Topics here are covered in a running category.

When writing employee handbooks, I prefer to use an "individual topic" template that is very similar to how I write SOGs. I believe that this is more user friendly for policy changes and modifications. I include an issue date and a revision history (running list of dates when the policy has been changed). I also like to provide a policy purpose, scope, and any definitions. Lastly, I cover the details of the policy in a narrative format and include relevant attachments.

Figure 8–4 is an example of a handbook written using the individual topic format.

WHAT TO INCLUDE

One of the most challenging issues related to employee handbooks is knowing what policies to include. Like SOGs, although similar topics may be covered by all employers, not all policies apply to every organization. As an example, a volunteer organization may be able to omit information related to wage and salary administration but would include participation criteria related to training, meetings, events and possibly even call response. Similarly, a department with represented employees (unionized) may have a grievance procedure spelled out in the collective bargaining agreement and therefore can have that section omitted from the employee handbook. Just as I discussed when writing SOGs, the copy and paste, change-the-name process will typically not

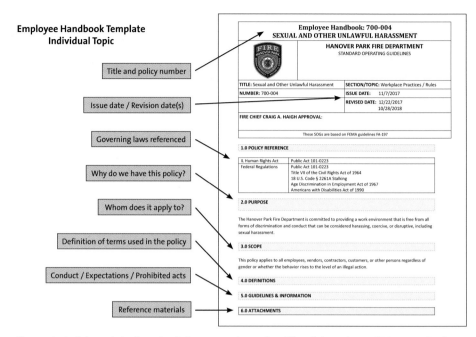

Figure 8–4. A template for a topic in an employee handbook (courtesy of Hanover Park Fire Department).

serve you well and is fraught with landmines. You need to do the hard work of researching and writing the policies. You also need to ensure compliance with all state and federal laws. Failure to correctly interpret and comply with these laws places your organization at great risk.

To provide some guidance on policy topics, several are listed in table 8–2. I have broken them into four broad categories. This listing is not all inclusive and is only provided as an example. It is imperative that each department/district research the standards applicable to their specific operation.

When doing the work of writing your handbook, each of these topics (if applicable to your organization) will need to be built out in a narrative format to specifically state how they will be handled by your organization. Although an employer has broad authority to govern workplace practices, many of these standards have very specific requirements within the law, and an organization would do well to research each and then have an overall legal review before implementing the policy. Speaking from experience, it is easy to miss stuff and find yourself in a legal battle trying to justify why you took the action you did.

Table 8–2. An example of policy topics.

Who we are as an organization	Workplace practices and rules	Employee benefits	Compliance with state and federal standards
Mission, vision, values	Work schedule	Timekeeping	Immigration law compliance
Chain of command	Safety in the workplace	Wage and salary administration	Family Medical Leave Act (FMLA)
Disclaimer regarding information within handbook	Appearance and dress	No pyramiding	Gift ban
Equal employment opportunity (EEO)	Computer use and social media policies	Fair Labor Standards Act (FLSA)	Court time
Nature of employment	Smoking policy	Leave time benefits: sick, vacation, personal, comp, holiday, professional development, funeral, and bereavement leave	Jury duty
References and background checks	Intoxicating substance usage		Time off to vote
Change in employment status, promotion, and demotion	Drug testing		Pregnancy leave and child bonding leave
Harassment, bullying, and discrimination	Discipline		Breast feeding
Code of ethics	Grievance process		Adoption leave
Employee conduct and practices	Evaluation and job performance review	Insurance: health, dental, vision, life, COBRA, workers comp, and disability	Military leave
Requirement to comply with internal investigations	Fitness for duty and medical exams		Family military leave
Employee complaint process	Inspection of personnel files	Pension	Victim's economic security and safety leave (VESSA)
Employee assistance program (EAP)	Workplace injuries and illnesses	Tuition reimbursement	Americans with Disabilities Act (ADA) compliance
Conflict of interest	Weather emergencies		Leave of absence
Maintenance of licenses and certifications	Secondary employment		Bone marrow donation leave
Separation from service and return of issued equipment, credentials, and keys			Organ donation leave
			Blood donation leave

ACKNOWLEDGMENT FORMS

A major part of both SOG and employee handbook administration is documenting that employees have received the policies along with any updates. It is imperative not only that the employee receive the documents but that they are also trained on the policies.

The Amboy (IL) Fire Protection District utilizes the forms shown in figure 8–5 to track acknowledgment of receipt of documents. Employees sign these documents, which are then maintained in their individual personnel files.

Until recently Hanover Park utilized a similar model. Employees would acknowledge receipt of the documents with training on the new and updated policies being conducted typically during morning roll call. Over the last two years the department has transitioned to a TargetSolutions software platform where each employee is sent an email notice that requires them to open the document and then acknowledge that they reviewed and understand the policy. This action serves to document that the employee has at a minimum reviewed the policy. New policies and changes are also still discussed during roll call.

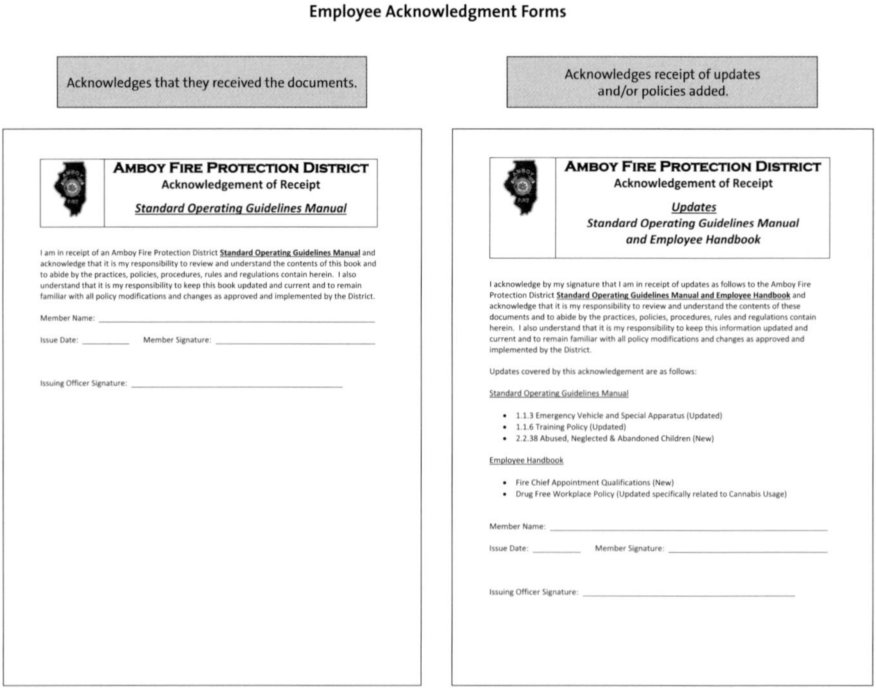

Employee Acknowledgment Forms

Acknowledges that they received the documents.

AMBOY FIRE PROTECTION DISTRICT
Acknowledgement of Receipt
Standard Operating Guidelines Manual

I am in receipt of an Amboy Fire Protection District **Standard Operating Guidelines Manual** and acknowledge that it is my responsibility to review and understand the contents of this book and to abide by the practices, policies, procedures, rules and regulations contain herein. I also understand that it is my responsibility to keep this book updated and current and to remain familiar with all policy modifications and changes as approved and implemented by the District.

Member Name: ____________________

Issue Date: ________ Member Signature: ____________________

Issuing Officer Signature: ____________________

Acknowledges receipt of updates and/or policies added.

AMBOY FIRE PROTECTION DISTRICT
Acknowledgement of Receipt
Updates
Standard Operating Guidelines Manual and Employee Handbook

I acknowledge by my signature that I am in receipt of updates as follows to the Amboy Fire Protection District **Standard Operating Guidelines Manual and Employee Handbook** and acknowledge that it is my responsibility to review and understand the contents of these documents and to abide by the practices, policies, procedures, rules and regulations contain herein. I also understand that it is my responsibility to keep this information updated and current and to remain familiar with all policy modifications and changes as approved and implemented by the District.

Updates covered by this acknowledgement are as follows:

Standard Operating Guidelines Manual

- 1.1.3 Emergency Vehicle and Special Apparatus (Updated)
- 1.1.6 Training Policy (Updated)
- 2.2.38 Abused, Neglected & Abandoned Children (New)

Employee Handbook

- Fire Chief Appointment Qualifications (New)
- Drug Free Workplace Policy (Updated specifically related to Cannabis Usage)

Member Name: ____________________

Issue Date: ________ Member Signature: ____________________

Issuing Officer Signature: ____________________

Figure 8–5. Employee acknowledgment forms (courtesy of Amboy Fire Protection District).

A WORD OF CAUTION

It has been the practice at Hanover Park for many years to give all new employees a CD containing department SOGs as well as the village's employee handbook as part of their rookie school training program. During the training program they instruct the new employees to review the documents on the CD, explain that they would be held accountable for the information, and then have them sign an acknowledgment form indicating that they have received the information. During rookie school, some but not all the policies are reviewed as part of the training program. As an organization we felt that we were in a good place related to how we managed our policy distribution to new employees.

That was until I had an arbitrator in an employee code of conduct case rule against me based on the fact that the information was "voluminous, containing several hundred pages" and that although we had records indicating that the CD was issued during the training period there was "no evidence that the Department explained or trained on any of these [code of conduct] policies and regulations." During testimony I explained our process for distributing the CDs as well as our process for handling policy changes and updates. I testified that we do not routinely review all policies with employees and that we have a level of expectation that employees will read the documents as instructed. I also testified that we do not routinely retrain employees on our policies unless there is a change. This caused the arbitrator to write, "Chief Haigh conceded that unless a policy is changed or updated, employees might not receive refresher training [on the individual policies]." Needless to say, this statement did not sit well with my bosses, and they were not pleased that I had missed what the arbitrator made to look like a serious misstep in policy management.

Based on this experience, I now recommend that departments/districts establish a routine where they review a group of policies with employees on regular intervals as part of a drill or roll call training. When this occurs, the training should be documented, including the names of those who attended the training.

CONTRACT POLICY DEVELOPMENT FIRMS

Due to the difficulties associated with writing SOGs and employee handbooks, I am seeing more and more departments utilizing consultants or firms such as Lexipol (https://www.lexipol.com/) to provide development and management of policies. Some of these firms offer online educational workforce training along with a continual review of the ever-changing state and federal standards related to employment law. For organizations that lack the internal expertise to research, develop, and maintain these critical documents, having a contractual or retainer relationship with a firm that specializes in this area is well worth the expense.

JOB PERFORMANCE REQUIREMENTS

JPRs are a standardized training plan for teaching and evaluating the skills we use on the job. When written correctly, they provide a step-by-step process that details how to perform a skill. Some skills lend themselves to a timed process. Others have a more "complete the step" focus. Some fall into both categories. As an example, the steps associated with starting a chainsaw are not necessarily time focused, while the donning of an SCBA has a specific set of steps that need to be accomplished within an established time standard. Writing JPRs requires analysis of the exact steps required to complete the skill. The author then needs to write them down in an easy-to-understand, step-by-step process.

Figure 8–6 is an example of a JPR utilized in Hanover Park related to the deployment of a cellar nozzle. The JPR provides an explanation of the objective, the testing standard, and the equipment needed to perform the skill. This JPR is a "complete all the steps" process based on the exactness of the skill, compared with others that may not require this level of meticulous performance. Because this is a required skill for our department, we regularly test firefighters on the topic and therefore have the name of the firefighter being evaluated at the end of the JPR as well as the evaluator's name and signature. Completed JPRs are forwarded to the training division for record keeping.

TIME AND MOTION STUDIES

Different from writing a "complete the step" JPR, when writing one that includes a timed component, the author needs to analyze both the steps as well as how long the process should take. When determining the time standard, the threshold should be based on the performance level of a reasonably trained firefighter. This means, do not set the time standard based on your star performer. It also means that you should not establish the time based on your weakest firefighter. The goal is not to set someone up for failure but to determine how long something should take and then train your personnel to perform within those established standards. To determine a specific time standard, a time and motion study needs to be conducted.

When I became a paid firefighter for the Rock Island (IL) Fire Department in 1988, they utilized what they called a "skills and maintenance" model that

Cellar / Bresnan Distributor Nozzle

OBJECTIVE: The firefighter, while wearing full structural firefighting equipment including SCBA, shall demonstrate the proper procedure for usage of the cellar/Bresnan Distributor nozzle.

Performance Standard: Accomplish the task with 100% accuracy.

Equipment Requirements:
- Cellar nozzle or cellar nozzle with shutoff valve
- Gate valve
- 50' 1¾" hoseline or 50' 2½" hoseline
- Sets of irons and splitting maul
- Pike pole/hook

Step No.	Required Action	Pass / Fail
1	The firefighter shall verbalize indications for usage of a cellar or Bresnan Distributor nozzle: • First handline has been ineffective. • Inability to reach the seat of the fire. • Inability to effectively ventilate. • Unusual layout of occupancy. • Inability to increase water flow. • Inability to increase access points to the seat of the fire.	
2	Firefighter will verbalize that an additional hoseline shall be in place for protection until the cellar nozzle is placed into operation.	
3	Check the stability of the floor.	
4	A hole shall be cut above the seat of the fire large enough to allow the cellar nozzle to be advanced through it.	
5	For 1¾" nozzle with shutoff valve: • Connect the nozzle to a 1¾" hoseline. • Push nozzle into ventilation hole. To properly operate, the nozzle needs to be suspended within the fire area. Typically, the nozzle should only be extended into the ventilation hole a few feet. • Open the bale of the cellar nozzle or shutoff. • Nozzle shall be operated at no more than 100psi. • Retreat to area of safety. • Monitor for changes of fire conditions.	

Figure 8–6a. An example of a job performance requirement, page 1 (courtesy of Hanover Park Fire Department).

	For 2½" nozzle without a shutoff valve: • Connect the nozzle to a section of 2½" hoseline. • At the opposite end of the hoseline from the nozzle place a gated wye 50' away from the nozzle. • Push both the nozzle and hoseline into the ventilation hole until the nozzle hits the floor below. • Nozzle shall be operated at no more than 100psi. • Open gated wye. • Move hoseline up and down in the ventilation hole to allow for maximum suppression of the fire. • Monitor for both changes in fire conditions and floor stability.	
6	The firefighter shall verbalize the limitations of the cellar nozzle: • Must be used above the seat of the fire. • Nozzle is used to "darken down" not extinguish the fire. • Operation above 100psi causes the nozzle to swing side to side, limiting extinguishing capability. • Reach of the nozzle is only 15'–20' from the ventilation hole. • Nozzle may have to be relocated depending on the layout of the fire floor.	

Name of Firefighter: _____   Date: __/__/__

Examiner (print): _____

Examiner (sign): _____   Date: __/__/__

Figure 8–6b. An example of a job performance requirement, page 2 (courtesy of Hanover Park Fire Department).

was based on time and motion studies implemented by the late Chief Glen Ayers. Time and motion studies are a way of analyzing specific tasks with the intent of finding the most efficient method of doing the work. This analysis is conducted using a standardized process with department-specific tools and equipment to determine the time required to perform a given task. Firefighters are then trained and evaluated on the standardized process.

Time and motion studies have been used successfully by businesses and industry for years. They are used in everything from assembly line manufacturing to fast food preparation. As an example, McDonald's Corporation, in training their crews, uses a training tool called Station Observation Checklists (SOCs). This is very similar to our JPRs and is based on time and motion studies. In an article for *Fire Chief* Magazine, Chief Ayers explained his thinking:

Why shouldn't a fire chief apply management procedures that business executives have found to be effective? Whether you are a manager of a fire department or a business, the principles of good management should apply. If a method works for business, we ought to at least see if it works for a fire department.

Time is the criterion. In business the time it takes to do a job has to do with costs and production. But in firefighting the time it takes to do a job may mean the difference between a life saved or a life lost. It could mean the difference between a large property loss or a small one. That's where the stopwatch comes in. (Ayers, 1969)

CONDUCTING THE STUDY

In developing time-motion studies, each step of the task is analyzed to determine what needs to be done, how the work is completed, and then how to make each step as efficient as possible. Take the pulling of an engine's crosslay hose load as an example. If I were to conduct this analysis, I would first ask a series of questions to ensure that the process we are using is the best one for our department. Sometimes we use a hose load because we have always used that load or because it fits well in the hosebed. I do not know that this justification always serves us well. I would argue that before we go to all the work of a time-motion study, we should first determine whether the hose load works well for the area we protect (fig. 8–7).

If we find something that works better, we should make the change before we conduct the study. Next, I would look at the steps needed for completion of the task and then lastly at how long it takes. I would work the process as follows:

- How is the hose currently loaded?
    - Question: In the respective response district for this particular engine company, does a need exist for the firefighter to be able to step away from the engine and immediately make a turn as they head toward the structure, or do you have room to clear the bed before needing to make a turn?
    - Question: Within the engine's respective response district, it is more efficient to carry hose or drag hose?
    - Question: Within the engine's respective response district, how much hose is required for the typical room and contents fire?

Figure 8–7. Hosebed of Hanover Park's Engine 16. In this photo personnel assigned to the company have expertly deployed a *Tactical Plan 2 Hose Lead-Out* for use at a structure fire using the standards established by the JPR (photo courtesy of Hanover Park Fire Department).

- Question: How is the crosslay hosebed constructed, and what type of hose loads can it accommodate?
- Question: Is the load deployable off either side of the engine or off one side only?
- Question: Based on the load utilized and the length of the hose—which is all driven by the typical demands of deployment—is the stretch designed as a one-firefighter deployment, or is a second firefighter required?

- What are the steps associated with deploying the hoseline? (For this example, I am using a 150' of 1¾" hose loaded in a Minute-Man configuration.):
  - Step #1: Grasp the nozzle and all hoseline flakes stacked on top of the nozzle. This should be 100' feet (two 50' sections).
  - Step #2: Shoulder load this stack of hoseline, paying attention to keeping the stack together and centering the load on the shoulder.
  - Step #3: Using the opposite hand from the one holding the shoulder load, turn and grasp the "deployment loop" of the remaining 50' section, pulling it clear of the bed.

- Step #4: Walk toward the fire, allowing the first 50' to deploy before you begin allowing hoseline flakes to deploy off your shoulder.
- Step #5: When you get to the front door, you should still have around 50' remaining on your shoulder. Lay this remaining hose on the ground so it can be charged with minimal kinks.

• Assess efficiency and potential for errors.
  - What steps can be eliminated?
  - What steps can be combined?
  - What can we rearrange to enhance efficiency?
  - What can we simplify?

The desired result of this process is efficiency. Once you determine that you are using the correct process and you have each step detailed with exactness, you can then determine through repeated drills with numerous different firefighters the average time it takes for each step of the process. Putting all these times together gives you the time required to complete the overall skill. This then becomes the standard. According to Chief Ayers,

> You wind up with procedures that work just like football plays. Each [firefighter] learns the "plays" set up for [their] equipment. With each firefighter following a prescribed routine, the result is a smooth team action. (Ayers, 1969)

PUTTING IT ALL TOGETHER

The beauty of JPRs based on time and motion studies is the fact that they can be combined to create drills that align with procedures (i.e., plays) established within our SOGs (i.e., playbook). To accomplish this, I am a fan of using Rock Island's model of aligning these JPR-driven skills in a side-by-side fashion based on what each company member should be working to accomplish while other company members are working on their specific assigned tasks.

> The theory is that when all members work to accomplish their assigned tasks, the operation is one of smoothness with little wasted motion, no shouting of orders, no scrambling for hose or tools. The firefighters go about the routine with speed—but without haste—calmly and unruffled. (Ayers, 1969)

Figure 8–8 uses several Hanover Park JPRs, including a time and motion study, to form the basis for a hoseline deployment model that is a mainstay of operations. In this case it incorporates the task of the driver/engineer with that of the officer and firefighter, or as we call it—the backstep position. It also provides a quick reference for the specifics of the stretch, tactical considerations, and performance requirements. This information is helpful for evaluators as well as members who are studying prior to participating in the drill.

FINAL THOUGHTS

Written documents take the guesswork out of managing both the team and emergency incidents. They define our standard of performance and govern our decision-making. They ensure consistency and a level of excellence that is impossible without their existence. So why do so many departments struggle in the development and implementation of these documents? I believe the answer to this question is multifaceted. In one sense, many of us have never learned or been trained to create these types of documents, and second, they are incredibly time-consuming to research and construct. As the fire chief/CEO, my best advice is to create a plan and start working to develop these documents. They are challenging, so your first few attempts may not be perfect. That is okay—at least you are trying and you without a doubt will be learning.

I also recommend that you share the workload of creating these policies, both as a succession planning tool and as a time-management tool. Most five-bugle chiefs simply do not have the time to do this level of work without assistance. Reliance on your team, while providing a level of oversight and review, generates buy-in, ownership, and a rare learning opportunity that will pay dividends in both performance and knowledge. Constructing these documents is hard, but it is well worth the effort.

REFERENCES

Ayers, G. (1969, April). Time and Motion Studies. *Fire Chief Magazine*, 15–19.

Grant, R. W. (n.d.). Part III—Alan Brunacini—The Fire Ground Command. *Fire Away: Interviews with Fire Protection Leaders*. United States Fire Administration. https://www.usfa.fema.gov/downloads/library/transcript_brunacini3.pdf.

Hanover Park Fire Department. (2003 & rev. 2007, 2012, 2017). Emergency Responder Rehab. In *Hanover Park Fire Department—Standard Operating Guidelines*. Hanover Park, IL: Village of Hanover Park.

Hose Lead-Out
Tactical Plan 2

Definition: A Tactical Plan 2 is a 2½" hoseline connected to 300' of 3" (making the available length of hose 600'). This line has a solid stream shutoff pipe connected to the male end of the 2½" hoseline. 100' of 1¾" hose is attached to the shutoff pipe with an adjustable stream nozzle (or smoothbore based on officer selection). The 1¾" hose is bundled together using straps.

Tactical Considerations: A Tactical Plan 2 allows water application with a line capable of flowing 180–200 GPM. The attack line can be rapidly converted to a single heavy stream hand line capable of applying 320 GPM for rapid knock down should the situation require. The playpipe should be located as close as advantageously possible to the area of involvement with enough 2½" line available to allow advancement to the seat of fire if necessary. Firefighters should expect to use this plan in any heavily involved structure fire.

This lead-out will allow truck and squad companies to have immediate access to the building (the front), making their operations in support of the engines attack more efficient.

Performance Requirements:
- Conducted with a three-person company (Driver/Engineer, Officer, Backstep)
- Calculate the length of the stretch
    - Building Width + Depth + Setback + ↑↓ floors = Stretch
    - Example:
        - 75' + 30' + 75' + second floor fire = 300' (6 lengths of 2½" hose)
- Evolution standards:
    - 200' of 2½" with playpipe and 1¾" positioned near the front door.
    - Conduct a reverse lay to the water source of 150'
    - Deliver tank water at proper nozzle pressure: 2 min. / 30 sec.
- **Evolution concludes when engine is on secure water.**

Total Evolution Time Standard: 4 min. / 15 sec.

Driver/Engineer	Officer	Backstep
Drive apparatus to the incident and position the tailboard so it is lined up with the hose deployment point.	While responding monitor radio and MDC for updates and check hydrant information and any available preplans. Provide safety message or reminders to company personnel.	Don SCBA, switch portable radio to fireground frequency.
	Don SCBA, switch portable radio to fireground frequency, prepare tools that will be used by the officer as appropriate (e.g., TIC, Halligan Bar, hand light)	
	Conduct size-up and transmit information related to findings, including the size of the structure. Provide direction of travel. Communicate "Tactical Plan" and	

Figure 8–8a. Hose lead-out tactical plan, page 1 (courtesy of Hanover Park Fire Department).

	provide any needed direction to incoming companies.	
	Assist driver/engineer with positioning tailboard of apparatus for maximum efficiency in conducting lead-out.	
Shift the transmission into neutral and set the parking brake.	Give order to dismount apparatus. Walk to the rear of the apparatus while calculating the length of the stretch.	Dismount when directed by the officer and report to the rear of the apparatus.
Dismount the apparatus and prepare to assist with the lead-out stretch.	Advise the driver/engineer and the backstep firefighter of the amount of hose you will be deploying.	Confirm with the officer the "Tactical Plan" that has been ordered.
When enough hoseline is deployed per the mathematically calculated stretch, return to the driver's seat and prepare to conduct a reverse lay to the water source.	Assisted by the driver/engineer, plan deployment of enough 2½" to cover the entire calculated stretch. In coordination with the backstep begin advancing the hoseline toward the structure.	Shoulder load the 100' of 1¾" hose as well as the playpipe and step no more than 3' away from the apparatus tailboard. Wait for directions from the officer before beginning the advance.
After positioning at the water source, again place the transmission into neutral, set the parking brake, place the pump into "pump gear," and set the transmission into drive.	Flake out the 2½" hoseline in a fashion to minimize kinks and allow easy advancement. If front door is standing open, work to close the door until you have water and are ready to begin the push.	Position 1¾" hose on the ground in a zigzag fashion and unbuckle the straps holding the load together. Position the playpipe as close to the entrance door as possible.
Dismount and connect the attack line to the appropriate pump discharge and send tank water. Use the hose chart to determine pump pressure based on the length of the lead-out and which nozzle is being used.	Check the backstep firefighter to ensure all PPE is appropriate for the fire attack.	Flake out 1¾" hoseline in a fashion to minimize kinks and allow easy advancement.
Prior to connecting to a hydrant, flush the hydrant to remove any debris collected inside the barrel. Connect the 5" soft sleeve to the hydrant and then to either the front suction or the valve connected to the eye of the pump.	Give direction to charge the 1¾" hoseline from playpipe. Open door and begin advance/fire attack.	Charge 1¾" by opening playpipe based on officer's command. Bleed air from 1¾" line by opening nozzle and pointing toward the ground. Begin advance/fire attack at the direction of the officer

Company/Officer Name: _____   Date: ___/___/___

Examiner (print): _____

Examiner (sign): _____   Date: ___/___/___

Figure 8–8b. Hose lead-out tactical plan, page 2 (courtesy of Hanover Park Fire Department).

QUESTIONS FOR FURTHER RESEARCH, THOUGHT, AND DISCUSSION

1. What is your organization's protocol for maintaining historical copies of policies (SOGs, JPRs, employee handbooks)? Based on applicable statute of limitation laws, how long do each of these need to be maintained?

   _____
   _____
   _____
   _____

2. Review your organization's employee handbook. Is it written in the "running category format" or the "individual topic format?" What is the most current review date?

   _____
   _____
   _____
   _____
   _____

   a. Based on your review, list five topic categories that need to be added or deleted based on current workplace practices and standards:

      i. _____
      ii. _____
      iii. _____
      iv. _____
      v. _____

   b. Based on your review, list five areas where information needs to be expanded to provide a more holistic overview of practices or to reflect changes to applicable laws and legal decisions.

      i. _____
      ii. _____
      iii. _____
      iv. _____
      v. _____

3. Review your organization's standard operating guidelines. Based on your review, list five guidelines that need to be added/modified/updated based on current operational practices. Describe in a general fashion what needs to be done for each.

   a. Guideline/policy: _____
      i. Needs: _____
      ii. Needs: _____
      iii. Needs: _____
      iv. Needs: _____
      v. Needs: _____

   b. Guideline/policy: _____
      i. Needs: _____
      ii. Needs: _____
      iii. Needs: _____
      iv. Needs: _____
      v. Needs: _____

   c. Guideline/policy: _____
      i. Needs: _____
      ii. Needs: _____
      iii. Needs: _____
      iv. Needs: _____
      v. Needs: _____

   d. Guideline/policy: _____
      i. Needs: _____
      ii. Needs: _____
      iii. Needs: _____
      iv. Needs: _____
      v. Needs: _____

   e. Guideline/policy: _____
      i. Needs: _____
      ii. Needs: _____
      iii. Needs: _____
      iv. Needs: _____
      v. Needs: _____

4. What is your organization's practice related to employee policy and SOG retraining? How would you strengthen the current retraining practice?

_____
_____
_____
_____
_____

5. Pick a JPR skill used by your organization. Using the following chart, write a step-by-step process detailing how it is to be performed. Provide an explanation of the objective, or why a firefighter would need to know how to perform this skill, and the performance standard against which the skill will be measured. Also list all equipment needed to perform the skill.

JPR SKILL

OBJECTIVE:

Performance Standard:

Equipment Requirements:

Step No.	Required Action	Performance Time Standard (if applicable)	Pass / Fail
1			
2			
3			
4			
5			
6			
7			
8			
9			
10			

Name of Firefighter: _____ Date: ___/___/___

Examiner (print): _____

Examiner (sign): _____ Date: ___/___/___

CHAPTER 9

RELATIONSHIPS MATTER

While speaking at a conference a few years ago, I was asked by one of the participants how I have had such a long tenure as a fire chief. The question frankly threw me, and I was not prepared to give an answer. I had not previously given it much thought. I just do—what I do.

Fortunately, my wife, Beth, was with me, and seeing that I was struggling to answer, she jumped in to provide some commentary. She said that I was the type of leader whom people just want to follow. They see my passion and vision for excellence and that this becomes infectious. She said that my style of communication makes my vision relatable and helps the team members see the big picture. Most importantly, they trust me. Honestly, I was a little embarrassed by her description and accolades, but both the question and her comments started me thinking.

1. Why have I been able to do the hard work and weather the inevitable storms of working as a fire chief/CEO for more than 30 years?
2. Why have I been able to continually move my organizations forward, while managing the multiple personnel issues involved in leading a department and at the same time not allowing myself to disconnect from the team?

I first became fire chief in 1991 when I was asked to lead my hometown volunteer department. I assumed this volunteer position while I worked full time for the City of Rock Island Fire Department. This volunteer position coupled with my full-time job and my teaching at the Illinois Fire Service Institute (IFSI) opened the door for me to eventually move to King (NC) as their first full-time chief (1995) and then on to Hanover Park (IL) as their fire chief (2002). All this adds up to me wearing five bugles (as either a volunteer or full-time chief) for more than three decades.

So, the question—is there a secret to my long tenure? Based on my experience, the best answer I can give is yes—I believe it is all about relationships!

RELATIONSHIPS AND PROBLEMS

James Flint, while serving as city manager in Alameda, California, wrote,

> Effective organizations are based upon effective relationships, and these are based upon trust. You can only build one trust-based relationship at a time. But this kind of partnership can multiply more rapidly if an organization behaves in accordance with certain shared values, focusing on honesty, respect, empowerment, collaboration, transparent behavior and open communication. (Flint, 2002)

Relationships are essential, but I would go one step further related to relationships and the leader. Team members need their leaders to be problem solvers. In order to do that, there needs to be a relationship where the team member feels comfortable bringing a problem to their leader. Relationships at this level are different from a friendship relationship due to the manager/subordinate component. Rather, these relationships are based on the "trusted leader relationship" whereby team members know they can depend on the leader to help them solve problems.

I am reminded of a quote by the late Colin Powell:

> Leadership is solving problems. The day soldiers stop bringing you their problems is the day you have stopped leading them. They have either lost confidence that you can help or concluded you do not care. Either case is a failure of leadership. (Powell, 1995, p. 52)

In order to get employees to bring you their problems, you must have a relationship with them. They need to trust that you care, that you will take the time to listen and to help, that your advice is sound, and that you won't hurt them, in either your solutions or because they showed vulnerability and asked for assistance. I try very hard to be the leader who gets the call when someone needs to find a solution to a challenging situation, and I try hard to always help, not hurt.

Besides talking to my own team members, one of the things that I enjoy the most is getting to interact with personnel from other fire service agencies and organizations. I routinely field phone calls and emails asking for advice based on my experience. These contacts come from people whom I have met along life's travels and with whom I have made a connection. I consider it as a great honor that I am the one they decide to contact for help. I also realize that their trust in me and their willingness to be vulnerable requires me to proceed carefully and to make sure that I truly help and not hurt the situation. I take it as a sacred responsibility. I also take it as a unique opportunity for me to learn and sharpen my own skills. This learning then translates into me doing a better job in leading my own team. It is a pretty amazing thing how these relationships intertwine and interact, allowing all involved to grow and develop. The importance of having this network of counselors is discussed more in the next chapter.

RELATIONSHIPS AND THE TENURE OF TODAY'S FIRE CHIEF/CEO

I am not aware of a study that has looked at the average tenure of today's fire chiefs. Based on my years of experience, I believe that it is somewhere around four to five years. This is not a scientific analysis but simply my gut read that comes from watching different departments across the country. I think that this low tenure is driven by a couple of different scenarios:

1. Fire chiefs/CEOs are often appointed late in their careers, and simply due to the chief's age and years on the job when appointed, they don't tend to stay long prior to retirement.
2. The four-year election cycles of mayor/village president/township supervisor/county boards also play a role in fire chief/CEO tenure. Since fire chiefs are part of senior staff/department head/cabinet level positions, they tend to ebb and flow with the election cycles.
3. Even if the political climate is stable, chiefs often wear down quickly, and their stamina diminishes as they work to steer the direction of their organizations. Often this occurs due to the intense pressure applied to this singular person from both managers/mayors/boards as well as the fire department personnel themselves (e.g., membership pressures, union pressures). These forces when applied together

suffocate the leader. In a phrase, the weight of the fifth bugle becomes simply too great.

This limited tenure of fire chiefs/CEOs is challenging and can be problematic for a few reasons:

1. It takes time to learn the job of fire chief/CEO and to garner enough experiences to be able to discern the best course of action for any number of challenging situations. I think that one of the primary reasons that other leaders seek my advice is because of my long tenure as fire chief/CEO. I liken it to the experience gained by going to emergency incidents. The more responses you make, the more knowledge and varied experiences you will have. The goal is to learn and grow from those experiences. The same is true for the fire chief/CEO. The longer you serve in the top spot, the more experiences you are going to have—many of these experiences are under extreme pressure. Short tenures limit your personal growth and development.
2. Any large project or significant change in how the department operates takes time to complete and implement. The larger the department, the longer it takes. If you are trying to change the culture of the organization, this simply does not happen during a short tenure. Firefighters are extremely resilient. If they don't like or believe in the cultural change you as the fire chief/CEO are trying to implement, they will either simply wait you out or will work to shorten your tenure. Either way, your overall influence is dramatically reduced. I remember well the words of Trustee Wes Eby, who was a member of the village board that hired me in Hanover Park. Trustee Eby was a "Navy Man" and often spoke of his experiences serving our country. He told me that he viewed changing culture like trying to turn an aircraft carrier around: "It takes a little time if you don't want the heel to be so severe that you start dumping multimillion-dollar aircrafts into the drink." The translation—if you want to effect real change, it is going to take a while, so commit for the long-haul.
3. Short tenures don't allow the fire chief/CEO to build political capital. Relationships take time to develop. Trust does not come overnight. You need to gather a few wins before you can count on receiving the benefit of the doubt. Your bosses, the community, and your team all need to see your performance over time. They need to see if you are who you say you are, even when it gets hard.

Lorri Freifeld, editor in chief at Lakewood Media Group, LLC, wrote, "One of the most profound experiences we can have in our lives is the connection we have with other human beings" (Freifeld & Webb, 2013). In the case of the fire chief/CEO, the connection with your team is of paramount importance. That being said, I also believe that your team needs to understand two distinct principles related to your relationship:

1. First, you will always look out for the best interest of the organization. That is your job! This involves focusing on what is right for the constituents of the community along with the well-being and reputation of the department.
2. Second, they need to know that they are personally important to you and that you will do all possible to protect and support them—within limits. They need to understand that your protection and support cannot be unconditional when it comes in direct conflict with the best interest of the organization. You may care deeply for them as a person, but they need to understand that you will also carry out the responsibilities of the office of fire chief.

For the chiefs who have had long tenures in the top spot, I think that their success comes down to relationships.

THE ROLE OF RELATIONSHIPS AND POLITICAL CAPITAL

Political capital is a term for the accumulation of power built through relationships and trust (Kjaer, 2013). The future success of the fire chief/CEO is often predicated on political capital that is built based on past successes. I think of political capital as a type of savings account where goodwill is banked based on successes. Each time a fire chief/CEO delivers on a promise, fixes a problem, remains true to their word, or performs well in any number of other similar scenarios, goodwill (i.e., capital) is banked for future use. Political capital can be spent, lost, or saved based on how the holder decides to use it.

It is also important to note that political capital does not only apply within the constraints of manager and board relationships. Political capital savings accounts exist for the fire chief/CEO in a variety of different accounts in a variety of different banks (e.g., the city manager bank, the city council bank, the union bank, the department membership bank, the neighboring department/mutual aid bank) Spending/investing from one savings account may generate increased political capital in one or more of these banks while simultaneously decreasing the capital available in another (fig. 9–1).

Figure 9–1. Political capital exists in a variety of different banks.

As an example, the fire chief/CEO may go to bat with the city council over an initiative that is important to the fire department union members. This initiative will likely increase the chief's political capital with the union and department membership while spending some of the capital accumulated with the council and manager. In these cases, it is imperative that the fire chief/CEO constantly be monitoring savings account balances to determine whether sufficient levels of goodwill exist before they decide to make a withdrawal to accomplish a potentially controversial goal or objective. If enough capital does not exist in the appropriate savings account, you need to wait until you have amassed enough resources to pay for the initiative. Timing is everything.

The reality is that for a fire chief/CEO to survive in this position, they must not overdraft any one of the various bank accounts that hold their political capital. When this occurs, most times you end up looking for a new job.

Elected officials, governing boards, and city/county/township managers typically work well with fire service CEOs who have built strong relationships with them and who have been successful with past projects and initiatives. This success builds political capital that tends to foster enhanced relationships.

These relationships are strongest when they are based on honesty, a record of consistently doing what is right for the constituents, and having shown that you are a team player who understands the various perspectives and the big picture agenda of the community. It takes time to build relationships, and in turn, political capital.

CASE STUDY: NEW BROOM MANAGEMENT THEORY

Let's use a hypothetical situation to explain how political capital is built and used and how it is related to timed expenditures.

Let's say for example that you have been hired as the fire chief/CEO from outside the organization. You are here because the majority of elected officials and the city manager believe that the department is having problems that can only be corrected through the "new broom sweeps clean" theory of management. This theory states that a new outside leader is needed to sweep away the stuck-down debris, dust, and dirt in order to truly clean things up and get the organization headed in the right direction. You, my friend, in this hypothetical scenario, are the new broom!

As you begin to assess your new organization, you quickly find that you have inherited a member of your executive team who appears to be more problem than help. This member seemingly is not interested in working with you to bring about change. You quickly have come to believe that he will boldly work against you, undermine your authority, cause disharmony, and will make things difficult as you work to accomplish the marching orders you have been given by your boss. It is apparent that this individual wields considerable influence within the organization, and he has a substantial following.

Compounding this challenge is the fact that he is personal friends with several city council members. You have been told by several people that he is dangerous and that you should be careful. You start to think that he may be the proverbial school yard bully, and you are seeing that many within the department seem to be afraid of him. He wields power using his rank as an officer and his strong political connections to intimidate and control the membership.

Strangely though, the elected officials with whom he is friends see something very different than what you have found in your quick assessment. They believe that he is a trusted member of the department who is only interested in doing what is best for the organization and the city. They believe that he works extremely hard and is the one who truly knows what is needed to right the listing ship. They don't perceive his actions as that of a bully but rather of a strong, out-front leader who doesn't mince words in his drive to get things done. In fact, if they could have swung a majority vote of the city council, they would have made him the new fire chief instead of you. Really, it was only the city manager who stood in the way, insisting that they hire a chief from the outside. Welcome to the new job, Chief!

Here are the challenges and a few options from my perspective on how to handle this type of situation:

1. Due to the political connections (relationships) and your short tenure within the city, you as the newcomer have established little to no political capital. Therefore, trying to take on this powerhouse within the first 100 days is likely a very poor decision.
2. You could try and win him over to your side. You need to figure out whose perspective of this individual is right—is he the school yard bully or the true leader whom the city council sees? If he would be willing to join you in trying to make the department stronger, you might find an ally in moving your agenda forward. I like this tactic best and recommend starting here.
3. If you are unable to win him over and you have determined that he is a cancer within the organization, your next best move is to minimize his impact. Think carefully about how this can be done without sparking a war within the department and city.
4. The true power for minimizing his impact comes from within the other members of the organization. If you can convince a few of them to follow your lead and give change a chance, the successes will embolden their courage, getting them to take a stance for what is right. It is difficult, even for the bully, to stand strong when the rest of the members are contrary to his positions.
5. While this is occurring, you need to begin achieving some wins with the manager and city council. These wins (even the small ones) build your political capital.
6. In the end, if you can't get your problem employee to come around and it has become obvious that he is the roadblock to doing what is required, you need to assess whether or not you have amassed a large enough amount of goodwill in your political capital account to go after him. In short, you need to have more political capital in your bank account than he does. If he has more than you, you lose. When you lose in this type of battle, in most cases it is the beginning of the end for your tenure within this organization.

Using a personal example from one of my commands, I waited four years to build enough political capital to push out a senior member of my command staff, even though I recognized within days of my arrival that he was going to be a problem. I tried to bring him along, but it was obvious that he was not going to be led by me. The lesson I learned—don't jump too fast, and always try to find other solutions before you make the decision to go after someone of great influence. You may win the battle but ultimately lose the war. Remember, you don't have to look far to find a pile of former, now unemployed fire chiefs who made the mistake of moving too fast and taking on the wrong person before they had enough political capital to not bankrupt themselves during the fight. Proceed carefully . . .

RELATIONSHIP WITH THE BOSS

We all have a boss. Some fire chiefs/CEOs report to a village/city/township/county manager (council-manager form of government), while others will report directly to a mayor/township supervisor/board president (strong-mayor form of government). Fire chiefs/CEOs who work for fire districts typically report to the district board in a similar fashion to the council-manager form of government. Regardless of the reporting structure, as the fire chief, you need to work hard at developing a relationship with your direct report.

In my case, I report to the village manager, who reports to the village board. This is the reporting structure that was used when I worked for the city of King (NC) as well as in Hanover Park (IL).

Village/city/township/county managers are responsible to oversee all facets of governmental operations within a municipality, township, or county. They utilize their department heads (e.g., fire chiefs, police chiefs, public works directors, park and recreation directors) to oversee the various departments and to manage operations within these specific organizations. It has been my experience that some managers are very hands-on and like to be involved and be made aware of every decision that occurs within a department, while others are much more focused on the big picture and give their department heads broad authority and latitude to manage their organizations.

Managers come from all different backgrounds and experiences. Many managers went to "city manager school" and graduated with a degree in public administration or similar with an ultimate plan to work in this type of position. Others served as department heads and found their way to the manager's office through experience and promotion. Occasionally you will find a manager who transitioned from the private sector or the military. Regardless of their background, the fire chief needs to learn to work with their direct report if they want to be successful in leading the organization.

A challenge with managers is that they often have short tenures within the political environments of communities. Elected officials change, and new agendas are brought to a community with each passing election cycle. Managers are responsible to get things done on behalf of the elected officials, but often what the elected boards want cannot be accomplished with the available revenue streams or governing statutes. This puts the manager in a position where they need to tell the elected officials that they can't do what they want or what they promised to do during their campaigns. Running a governmental organization takes money, and if you want to accomplish a bunch of initiatives, you need to have a bunch of money. Money for initiatives in government comes primarily from taxes, and no elected official wants to raise taxes (i.e., require their constituents who voted them into office to pay more). This then becomes a paradoxical challenge dumped at the feet of the manager, making their job very difficult and, in some cases, impossible.

The volatility of managers caused by the challenges of the election cycles often creates difficulties for the fire chief. As I discussed earlier, many initiatives in the fire service take many years to accomplish and involve significant planning and hard work. Each time you have a manager turnover, the fire chief needs to again build a relationship with the new boss. Part of this relationship building typically involves rejustifying how you operate, including the initiatives that you have underway. Sometimes based on manager changes the priorities and initiatives get changed, which forces the fire chief to modify the direction of the organization. Unfortunately for many fire chiefs, due to these situations, they find themselves in a shifting workplace environment that is often untenable. Due to these demands, or in some cases the new manager's leadership style, the tenure of the fire chief is often shortened.

What I have found that works well for me to maintain communication with my boss is a biweekly written report detailing the activities of the department. This report covers in a narrative bullet-point fashion the work that has been accomplished, challenges that have been encountered, and an overall review of number matrixes related to department operations. These reports build off the one before and are designed to show progress and to clearly paint a picture of the status of the department. (See the section on check-in meetings for bosses in chapter 6.) It is easy in our fast-paced world to miss something during a conversation with our boss, but when you take the time to write it down, I find that you are able to more clearly address issues and ensure that the manager is kept informed. This document is a key in the overall transparency of department operations and management in that it is informative, show trends, and does not hide issues of concern. Once the manager receives this document, if they have questions, they can follow-up as appropriate.

My personal philosophy is to not bring issues to the manager without a recommended solution or action. It is my responsibility to be the expert in fire department operations. This means that I need to do my research, consult with my leadership team, and work to figure out a plan instead of just dumping a situation on the manager's desk and waiting for instruction. Sometimes, however, your recommendations cost money, are difficult to implement, or involve personnel and personalities. In these cases, the conversations, although needed, can be challenging.

Challenging conversations with the manager require a relationship where you both can be honest. Sometimes what you have to say is difficult, but keeping it hidden does not help the organization and certainly does not keep the manager informed. It has been my experience that some managers do better with hard conversations than others. Having a good relationship with the manger sometimes helps—but not always. Hard conversations are . . . just hard.

Also, it is your responsibility to not allow your boss to get blindsided. If you see something coming down the road or you know of a situation that is about to unfold, it is your responsibility to bring this forward. This includes conversations you may have had with elected officials. The partnership between you and the manager must be considered one of the fire chief's/CEO's essential relationships.

Lastly, do not do an end run around your boss to circumvent their authority by appealing to a different authority (i.e., the elected board). You need to work through the appointed structure of your organization and follow the chain of command. Just like at the fire department, sometimes you make decisions that are controversial and not supported by everyone. You make these decisions based on what you believe is the right direction for your organization using information and facts that you have available to support your position. Fire department personnel should not jump over you and bring their appeal to the manager. Similarly, the manager will sometimes make decisions that you don't agree with or support. When this occurs, you should not run to the elected officials. Even in situations where you have a serious ethics issue with the manager, you need to first bring your concerns to the boss, giving them the chance to address the concern. Rarely if ever should you assume the whistle blower role as your first step in addressing a problem. It has been my experience that honest conversations usually fix most issues. But to have these hard conversations, a relationship needs to exist.

I try to follow the biblical standard outlined in Matthew 18:15–17 related to these types of issues:

1. Go to your boss first and explain the situation. Give them the chance to make corrections.
2. If they don't listen, bring a couple of witnesses so the matter may be established by the testimony of others. Bringing witnesses means other team members who have knowledge of what is happening and who can support the argument you are making through their own personal testimony, using facts and situational evidence.
3. If they refuse to listen to the witnesses, then and only then is it okay to elevate a concern.

If the manager is worth their salt, they will listen and make corrections.

RELATIONSHIP WITH THE ELECTED BOARD

Often, due to the nature of the position of fire chief, elected officials will reach out to you directly to answer a question or find out information. Although this is acceptable and probably should be encouraged, if you work within the council-manager form of government, when it happens it is wise to drop the manager a note to let them know about the conversation. Sometimes this notification is nothing more than a heads-up. However, you need to be cautious in that you may be being used by the elected official to push an agenda that you are not aware even exists. Also, sometimes you will have an elected official who is trying to work around the manager and is using you as the pawn. Be cautious in these situations and make sure that you notify the manager when they reach out to you. This will avoid problems down the road.

It is also important to understand the idiosyncrasies of the community in which you work. As an example, in Hanover Park the village president (he likes to be called mayor) is hugely engaged in the leadership of the village far beyond just chairing a couple of meetings each month. He basically works full time and keeps regular office hours. He is an astute politician who is unrelenting when it comes to taking care of his community. He is the kind of leader who is unafraid to walk into the office of any legislator, whether at the local, county, state, or federal level, and demand an audience about what is needed in his community (fig. 9–2). He spends time in the schools each week reading to the students. He regularly stops in to businesses to talk with the owners and to thank them for choosing to locate in Hanover Park. He is on several

Figure 9–2. Hanover Park's mayor, Hon. Rodney S. Craig, Chief Haigh, and Illinois House Representative Fred Crespo at a fire department open house event (photo courtesy of Hanover Park Fire Department).

boards and can regularly be found about town talking with residents and being engaged with the local Veterans of Foreign Wars (VFW) Post. He is a powerhouse of political clout and is known for his ability to get things done. I would also say he is loyal to the core!

Although the organizational structure shows that I reported to the village manager, the mayor was a constant presence, and because of his leadership style, I felt that I was responsible to him as well. This was not always an easy and clean reporting structure in that the mayor would call or stop by my office, give me direction, and then move to his next engagement expecting me to let the manager know what he has just decreed. Sometimes this caused challenges, but I will say that I truly loved working for him.

In many ways the mayor leads the village as if we operate under the strong-mayor form of government, but that is not our system, and all of us in leadership have learned to work with the manager as well as the mayor in handling the affairs of the village. This is our village's culture.

Fire chiefs need to learn the culture of their communities and then figure out how to work within the leadership structure as it exists. Successful and long tenured chiefs have figured out these relationships and how to make them work. This is a critical part of the job.

Lastly, when working with elected boards in the council-manager form of government, it is important to remember that it takes a majority vote to direct something to be done. Therefore, just because an elected official gives direction does not mean it is a dictate unless it is supported by the majority of the elected body. I have watched elected officials over the years place huge demands on managers and department heads without the majority support of the elected body. As the fire chief/CEO you need to be judicious in saying no (sometimes it is just easier to comply), but you also need to be prepared, if the demand is too great or you believe it may be controversial, to request that the elected official seek a majority vote of the board to direct you to do the work. Otherwise, you often find yourself spending all your time supporting studies and projects of the elected officials that may or may not be important to the whole community.

But again, you also need to work within the culture of your community. In Hanover Park, if the mayor requests something to be done, it is wise to simply do it. The elected body gives him broad authority beyond what would be normal and customary with a village president. The manager and department heads understand this cultural issue and work to support the system as it exists.

Also, any information shared with one elected official needs to be shared with all members of the elected body. As an example, if one elected official requests the fire chief to provide data on the number of drug-induced overdoses within the community over the last three years, this may be an acceptable request depending on the work associated with compiling this information. Once you have the report complete, it needs to be shared with all elected officials and not just the one making the request. This avoids or minimizes the political "I got information you don't have" fights that often occur within political bodies and avoids putting you as the fire chief/CEO in the middle of these internal power struggles between elected officials.

RELATIONSHIP WITH THE TROOPS

The fire chief/CEO cannot move an organization forward by always fighting with the troops. Just like the relationship needed between the fire chief/CEO and their boss, chiefs must also continually stay connected to the internal environment of the department. This connectedness comes through relationships with staff members at all levels of the organization. If I had to pick one of the primary reasons that fire chief/CEO careers unravel, I would say unequivocally that it comes from poor relationships with their personnel. Whether paid or volunteer, poor or strained relationships between the workers and management will cause significant challenges within an organization. Even though the fire chief/CEO is the player at the top, without some common alignment of vision and goals, as well as the general belief that the boss truly cares about the membership, things will go awry.

Chief Dennis Compton (Ret.) of the Mesa (AZ) Fire Department wrote, "The way the workforce and management interact has a direct impact on the quality of service the customers receive, [both] internally and externally" (Compton, 2003). The fire chief/CEO may head the organization, but it is the membership/employees who do the work of serving the community. They are the personnel on the front line who are tasked with solving problems for the residents, businesses, and visitors. Without them, the mission of the fire department does not get accomplished.

I still marvel at the sustained "them against us" mentality that exists in the fire service. This is not nearly as prevalent in the volunteer fire agencies, but it exists in overabundance for many (if not most) career departments. This mindset truly baffles me because we are all on the same team working to accomplish the same mission. Yet this antagonism is propagated at all levels of the organizations and is the fault of both firefighter and fire chief. The fact that we lose sight of this singular team truth is absolutely maddening.

The assumption that firefighters have that their leadership is plotting ways to "stick it to them" and who actually see "boogeymen in the dark" seems crazy when you look at it from the 30,000-foot view of the public eye. Yet there absolutely are fire service leaders who are truly the boogeymen who spend time dreaming up ways to mess with the team members. How much more productive could we be if we refocused that same time and energy on ways to better serve the public?

I would say that for many of us who serve in the top leadership level of today's fire service, there simply are not enough hours in the day to get the real work accomplished, let alone time to sit and scheme about ways to mess with staff. Frankly, we have way better stuff to do. This mindset and focus, on either side (labor or management), is just nuts in my book! Yet I will say that within my own organization, after years of trying to combat this issue, and although the water generally looks very calm on the surface, this boogeyman mentality still lurks just below and can jump up and bite when you least expect it.

So how do you combat this mentality? I think that it comes down to relationships and each side staying focused on their areas of responsibility and the overall mission of the organization.

Within states that allow collective bargaining, firefighters, through their union representatives, have the right to negotiate with management over wages, terms and conditions of employment, and employee safety practices. Management on the other hand has the right to determine the organization's mission, budget, and strategy and the operational right to assign, direct, hire, and discharge employees (discharge within the guidelines of agreed-upon discipline and discharge language within the collective bargaining agreement). This is the basic premise of collective bargaining.

It is my opinion that labor often oversteps its bounds related to management of the department/district. This occurs due to fire chiefs/CEOs who fail to take a strong stance as the leader of their organization and allow the union to take on roles that they have no business doing. The union does not have the right to manage the fire department/district. Their role is to protect the members. Yet you don't have to look very hard to find departments/districts that are run by the union where the fire chief/CEO is nothing more than a figurehead. Fire agencies need leadership, and in the absence of leadership from the top, labor has no choice but to grab the reins and do the job that should be done by the designated leader.

As the designated leader, it makes sense to manage the organization based on a team approach. The union represents the team members who do the work; therefore, working together based on the mission of the department tends to yield the best results. This is not abdicating the fire chief's/CEO's authority to the union but rather recognizing that leadership is leadership, regardless of which side of the bargaining table you sit on. Smart people working together to accomplish the same goal usually produces excellent outcomes. Therefore, harness the intellect of the leaders within the organization, regardless of whether they are formally designated as labor or management. Bottom line, internal organization strife does nothing but hinder and distract from the service we provide to our customers. Working together does not weaken the authority of the fire chief/CEO; rather, it strengthens it, as it allows you to make better decisions because you have insight gleaned from various perspectives and angles.

The International Association of Fire Chiefs teaches that there is an essential need for the creation of a leadership partnership between local fire chiefs and their union presidents. These partnerships need to be based on relationships. When egos get in the way of relationships, we begin to have challenges that are driven by hurt feelings. The late Chief Alan Brunacini of the Phoenix (AZ) Fire Department used to talk about two truths related to this very issue:

1. "Egos eat brain cells."
2. "The number one cause of firefighter line-of-duty injuries ... hurt feelings."

So what is the union really looking for from leadership?

- Fairness in how employees are treated
- Comparable wages and benefits
- A desire to be heard by management
- A desire to play an active role in deciding their future
- Safety and health of their personnel

Conversely, what is management really looking for?

- Ability to meet the mission of the organization
- Quality levels of service
- Minimal complaints from the public and the governing body
- Quality work force that is stable with low employee turnover
- Minimal personnel issues (e.g., discipline, injuries)
- Ability to pay

If we work to understand each side of the labor/management equation, it becomes easy to find common ground. This understanding, however, only comes through relationships that foster open, nonthreatening communications. Internal labor/management fighting does nothing more than take time and resources away from the real work that needs to be done. If you want to build quality relations, you must control egos, stop making threats, work to focus on the mission of the organization, and always lead with a sense of grace toward the other side.

My experience has been that frequent communication reduces formal procedures:

- *Advice to administration:* If you know a topic has the potential to cause concern or be controversial, bring it to the union up front and work through the issue before implementation.

- *Advice to the union:* Rather than waiting for administration to make a controversial decision or a mistake and then responding with a grievance, discuss the situation up front.

Share openly what issues you are facing. Working as a team will allow you to have a much better chance of finding the right solution for the global organization. I believe that a participative decision-making model not only increases employee productivity, job satisfaction, and employee development and growth, but it serves to develop your organization's formalized leaders. I find it fascinating that in Hanover Park, every International Association of Fire Fighters (IAFF) union president, excluding the one currently serving, has ultimately been promoted to a chief officer rank. The lesson is simple—leadership is leadership.

FIRST CALL RULE

Over the past 20 years, the Hanover Park Fire Department has enjoyed an unprecedented labor/management relationship. Working in the Chicago metro area with two different labor groups representing our employees (IAFF and Service Employees International), we have only had one grievance filed by either union in the past 20-plus years. This is a tremendous achievement on the part of many and one of my proudest accomplishments.

When I arrived in Hanover Park as the new fire chief, we were coming off a two-year stint where the IAFF local filed more than 25 grievances. The astonishing fact is that the union prevailed on each and every one. I knew that this could not continue, so I devised a plan and worked with union leadership to develop and implement what we have termed the "First Call Rule." This rule says the following:

1. If management notes a problem, the fire chief will place the "first call" to the union president to discuss what has happened. The chief will advise the president of exactly what he knows and will hold nothing back.
2. Likewise, if labor notes a problem, the union president will place the first call to the fire chief to discuss what has happened. Again, the president will advise the chief exactly what he knows and will hold nothing back.
3. A grievance will NEVER be filed unless both sides agree that it is the only course of action and no other options exist.

The key aspects of this rule are the following:

1. No one takes ANY action without a discussion.
2. There are NO secrets between the fire chief and union president.

As part of the labor/management relationship, we have developed rules of engagement shown in figure 9–3.

I think that our model of labor/management methodology is summed up well by the words of Chief Smokey Dyer (Ret.), Kansas City (MO) Fire Department:

> One of the critical success factors of every fire agency is the methodology for decision-making within the organization . . . when departments have most of their decisions made in the "front office" our personnel generally adopt an indifferent attitude and become a roadblock for positive change. Issues tend to be win/lose rather than what is best for the department or citizens we serve.

To successfully manage this win/lose mentality, relationships need to exist so that issues can be addressed and problems minimized in order to allow us to

HANOVER PARK FIRE DEPARTMENT

RULES OF ENGAGEMENT
LABOR / MANAGEMENT RELATIONS

- Open and honest communications (tell what you know).
- Be fair and consistent.
- Same rules apply to all.
- Don't make mountains out of mole hills.
- If you are wrong—admit it!
- If you make a mistake—fix it!
- Trust first!

Rules of Engagement—Fire Chief	Rules of Engagement—Union President
• Keep the team (especially the union president) in the loop as to what is happening within the organization. • Bad news needs to come directly from the fire chief.	• You have a "high security clearance"—understand confidentiality . . . • If you want to be part of the "inner circle," understand what that means . . .

Figure 9–3. Rules of engagement for labor and management relations (courtesy of Hanover Park Fire Department).

carry out, as a team, the overall mission of the organization. I believe the key to making it happen—it's all about building strong relationships.

REFERENCES

Compton, D. (2003). Labor-Managment Relations. *Firehouse.* https://www.firehouse.com/home/article/10528858/labormanagement-relations.

Flint, J. (2002). Mending Labor-Management Relationships. *ICMA.*

Freifeld, L., & Webb, L. (2013, March 21). 8 Tips for Developing Positive Relationships. *Training Magazine.* https://trainingmag.com/content/8-tips-developing-positive-relationships/.

Kjaer, U. (2013). Local Political Leadership: The Art of Circulating Political Capital. *Local Government Studies, 39*(2), 253–272. https://portal.findresearcher.sdu.dk/en/publications/local-political-leadership-the-art-of-circulating-political-capit.

Powell, C. (1995). *My American Journey.* Random House Publishing Group.

QUESTIONS FOR FURTHER RESEARCH, THOUGHT, AND DISCUSSION

1. Identify a person in your life with whom you have a "trusted professional" relationship. List five characteristics that they consistently exhibit that makes them a trusted ally.

   a. _____
   b. _____
   c. _____
   d. _____
   e. _____

2. Within your own organization, provide a listing of the top five "political capital" bank accounts maintained by the fire chief/CEO. List these accounts in order of importance along with a description justifying the ranking.

   a. _____
   b. _____
   c. _____
   d. _____
   e. _____

3. If you were to develop a regular written report designed to provide information to your boss so they are better informed related to organizational operations, functions, and needs, what broad topical categories would you include, and what statistical information would you consistently track and share?

   a. Categories:

      i. _____
      ii. _____
      iii. _____
      iv. _____
      v. _____

b. Statistics:

i. _____

ii. _____

iii. _____

iv. _____

v. _____

4. Describe the political culture of your organization related to elected officials.

   a. Reporting structure (formal and informal):

   _____
   _____
   _____
   _____

   b. Expectations:

   _____
   _____
   _____
   _____

   c. Style of management of the chief elected official:

   _____
   _____
   _____
   _____

   d. Potential pitfalls for the fire chief/CEO:

   _____
   _____
   _____
   _____

5. Conduct a SWOT Analysis assessing the current labor/management relationship within your organization:

   a. Strengths:

   _____
   _____
   _____
   _____

b. Weaknesses:

_____
_____
_____
_____

c. Opportunities:

_____
_____
_____
_____

d. Threats:

_____
_____
_____
_____

6. Evaluating your essential relationships, describe areas of current or potential concern along with three personal action steps that may help to address or minimize the challenges.

   a. Boss/direct report:

   _____
   _____
   _____

      i. Action step: _____
     ii. Action step: _____
    iii. Action step: _____

   b. Elected officials:

   _____
   _____
   _____

      i. Action step: _____
     ii. Action step: _____
    iii. Action step: _____

c. Command staff/officers/support staff (those you depend on daily)
   _____
   _____
   _____

   i. Action step: _____
   ii. Action step: _____
   iii. Action step: _____

d. Labor/membership:
   _____
   _____
   _____

   i. Action step: _____
   ii. Action step: _____
   iii. Action step: _____

e. Constituents:
   _____
   _____
   _____

   i. Action step: _____
   ii. Action step: _____
   iii. Action step: _____

CHAPTER 10

WISE COUNSEL AND THE WEIGHT OF THE FIFTH BUGLE

The year is 1992, and I am early into my role as chief of department for Hampton Fire Rescue. Hampton is a small river town with a population of about 2,000 in northwestern Illinois along the banks of the Mississippi River. It is part of the Quad Cities region and is primarily a bedroom community with a few businesses sprinkled along the main highway that parallels the river. Hampton is the community where I attended grade school and junior high, played sports at the county-owned ball diamonds, rode my bike, hiked the bluffs overlooking the river, and learned to drive a car and where I desired more than anything else to be a firefighter on that community's volunteer fire department (fig. 10–1).

I started hanging around the volunteer fire department as soon as they would let me in the door, and at age 16, they appointed me a "Reserve Firefighter." This reserve program pre-dates the explorer/cadet programs many departments operate today and was designed to spark the interest of young people to serve their communities through the fire service. The program was a huge success in that numerous members who joined as reserves went on to become career firefighters, fire officers, EMTs, paramedics, and registered nurses. Some remained as volunteers, while others left to follow their dreams as big city firefighters. Four of the early members of this program have climbed the job ladder to become chief officers in career departments, with two serving in roles wearing five bugles. Not bad for a volunteer department in a community of 2,000 people.

Figure 10–1. I always wanted to be a firefighter, especially a Hampton (IL) firefighter. I also dreamed of being Hampton's fire chief. For me, both dreams came true.

I am extremely proud of my heritage at this department. My reserve fire helmet continues to be prominently displayed in my office along with numerous pictures and plaques commemorating my time at Hampton (fig. 10–2). It is also fair to say that I will always be grateful to the place where I got my start in the fire service. Not just my start as a firefighter—but also as a fire chief.

I was hired as a career firefighter/paramedic with the city of Rock Island (IL) in 1988, not far from Hampton. This scenario was ideal for me; I got to learn, experience, and make a living as a career firefighter and then come home and continue to serve as a volunteer. In part because of this two-department situation, in December 1991 I was asked to serve as chief for my hometown volunteer department.

In this role, I learned early on that the chief's position is more than leading a group of friends to serve the people of a community you love; it is really the role of being the boss, and as President Harry Truman said, "The buck stops here" (Truman, 1952).

Figure 10–2. The author's first firefighting gear.

LEADERSHIP CHALLENGES

Volunteer departments are made up of members with diverse backgrounds and experiences. This in fact is what makes volunteer organizations strong, but it also sometimes invites leadership challenges. At Hampton we had a longtime member who was very dedicated and committed to the organization. He was my parents' age and had carried into adulthood some challenges in controlling his temper. He had on two prior occasions, once during my tenure as chief, either thrown a punch or threatened another member of the department. I had severely chastised him the last time it happened and had made it clear that this type of behavior would not stand under my command. But as fate would have it, it happened again.

We were holding our monthly volunteer firefighter's association meeting when he got into a disagreement with another member. I was sitting at the head table, which was customary for the fire chief, and before I knew what was happening, this individual, along with the one he was arguing with, both stood up, and my fighting friend delivered a well-placed kick to the groin of his surprised opponent. The room erupted in noise, and fortunately the two were quickly separated by firefighters sitting nearby. To say the least, I was caught off guard by this outburst of physical emotion, but I also quickly realized that the eyes of the entire department were watching me to see how I would handle this situation.

Here I am, a new fire chief in my mid-20s who had been appointed by the village board far earlier than was probably prudent, and I was now faced with a very visible disciplinary challenge from an individual who was old enough to be my father. For a few moments it seemed like my life was running in slow motion. I could see the two separated individuals in the back of the room, I could see many of the members looking directly at me, and oddly, the sounds of the room had become like dampened background noise and the voices indiscernible. My brain was processing what I had just witnessed, and I was trying desperately to develop an appropriate response. Suddenly the fog cleared, and I simply told him to leave, go cool down, and that he and I would talk later. I could see in his eyes that he knew he had done wrong, and without a word, he exited the meeting room.

Following the meeting I spoke to a few witnesses in the immediate area of the scuffle to get their perspective on what they saw and who was the first one to jump to their feet. I also spoke to the firefighter who had been struck. Although his face still had a slightly green color and he was walking hunched over, he was able to relay a story consistent with the witness statements. As I think back, I believe that this was my first official internal investigation.

Through this process it was quickly determined that my fighting friend was the primary aggressor and that his opponent had left his seat only as a defensive measure. Although disappointed in my fighting friend's behavior, I was also not surprised.

I remember calling him the next day and having a quick conversation where I said that he was being placed on leave while I considered my next steps. I admit, this was not some well-thought-out plan driven at the advice of my legal counsel (which is how I operate today) but rather a common-sense step to buy me some time to think and weigh out the best option, if one existed. Remember, this is a volunteer department, and without volunteers, calls don't get answered. It wasn't like I could just go and pull another firefighter from a hiring list. In most cases when you lose a volunteer member, you now just have a vacancy.

I thought about this situation day and night for the next two weeks. I fretted, lost sleep, prayed, talked with my wife, and tried to consider all my options. I felt completely alone with the full weight of the fifth bugle pressing down on my young shoulders. In my mind, I kept coming back to termination as the right option (remember, volunteer firefighters are employees of the organization and are subject to discipline up to and including termination), but then I would second-guess myself and lean toward showing grace and keeping a dedicated volunteer.

KITCHEN TABLE ADVICE

Finally, I decided that I needed some professional advice. I went to see Battalion Chief Robert "Bob" Collins, whose shift I was assigned to at Rock Island. Chief Collins was a wise and experienced command officer who had been a champion of me and my abilities since I had been hired by Rock Island. I believed that Chief Collins could provide sound advice and could help me figure out my next steps.

I approached Chief Collins after dinner while he and one of our senior firefighters played chess at the "Central House" kitchen table. He was well into his normal after dinner routine—a game of chess and a cup of coffee while his regular gaming partner munched on a bowl of freshly popped popcorn. I told Chief Collins that I needed his advice related to a situation at Hampton, and the duo immediately paused the game and I had his full attention. I explained the situation, what I had done, and the action I was contemplating. He asked a few questions and then quickly confirmed what I already knew: I had to let this individual go.

Chief Collins summed the situation up like this:

1. If I let him stay, it would send a message to the organization that this type of behavior is okay and that it would be tolerated under my command.
2. I had told this individual previously that I would not stand for this type of behavior in the future—he had been warned. Now, as Chief Collins advised, I needed to carry out what I had already promised that I would do—not tolerate this behavior.

After giving me this sage advice, Chief Collins paused, changed his demeanor, and asked me how I was doing?

I am sure that he could see the stress and fatigue in my face and eyes. He wanted to know how this situation made me feel and what concerned me the most. We talked about the best way to deliver the message to this long-tenured member and the importance of not letting the stress of being the boss erode my personal well-being. He acted as a confidant and mentor, and most importantly, it was obvious that he cared about me personally and about helping me do the best job possible. He was both providing emotional support and at the same time instructing me on what needed to be done, all the while sitting at probably the best classroom in the world—the firehouse kitchen table.

LESSON LEARNED

The action I needed to take was clear. If I didn't do what I said I was going to do, my word as a leader meant nothing. So I mustered my inner strength and told this firefighter that he could no longer be a part of the department. He turned in his issued equipment and through tear-filled eyes, handed me his badge and department credentials.

Here is the absolute truth . . .

In all the officer development classes I had attended, all the leadership books I had read, and all the one-on-one conversations with fire service leaders I had had, no one had ever explained to me the feeling of taking away a firefighter's badge.

When he handed me the badge, I literally could not speak. I simply looked at him for a long moment and then motioned with my head for him to leave my office. I watched through my office window as he walked to his car and drove away. I sat in my desk chair holding his badge while tears ran down my cheeks. My tears were not about the loss of this firefighter—I knew that I had made the right decision. The tears were caused from the crushing weight of responsibility and the realization of just how hard this job really is. How many more times would I be required to do this as a fire chief? Would it always feel the same? I sat in silence—alone—for a long time. Then I put the badge in my office safe and drove home.

I had just experienced for the first time three aspects that I have come to learn are constant facets of the job of fire chief/CEO:

1. The crushing weight of the fifth bugle
2. The loneliness of command
3. The need for wise counsel

COMMAND IS LONELY

The late Colin Powell, former secretary of state and a four-star U.S. Army general who served as the chairman of the joint chiefs wrote in his book, *My American Journey*, "Command *is* lonely" (Powell, 1995, p. 320).

Having responsibility for an organization, its team members, and the work they perform, while simultaneously setting the organization's vision and direction can be a crushing weight that can overpower even the strongest leaders.

As the fire chief/CEO you often feel alone, even when surrounded by key team members from your administration. These members, try as they might, cannot fully grasp the psychological weight of being responsible for all that occurs within the agency and through the delegated power to subordinates. Making decisions is an integral part of what you do, and sometimes your decisions carry a power that will have a negative impact on people's lives, take away their careers, and financially impact their families. Those factors all add to the weight of the decisions, but in the end, action needs to be taken, and the authority and responsibility rests with you.

It seems that somewhere in the excitement of the promotional process, we miss a very important key aspect of the job: great wisdom does not come naturally the moment you don the collar brass of a fire officer.

Great authority—yes

Great responsibility—yes

Great wisdom—not so much!

You can read books and articles, obtain formalized educational degrees, review case studies, and attend training lectures by world-renowned fire service leaders all to prepare you for the challenges of the job. You can hold meetings of your command team and legal advisors, ask insightful questions, and compile data on the likelihood of success based on various options, but in the end you have got to understand that the decisions you make related to your organization and your team members are completely your own. These decisions, when they involve a difficult topic, will without question impact your organization and your leadership legacy. You need to make them armed with the full picture of their potential impact, but most importantly, you need to make them!

WISE COUNSEL

One immensely helpful tool for leaders is to seek the wise counsel of those who have experienced similar situations and who can help you dissect the various aspects of a dilemma in order to make the best decisions. No matter the talent within your organization, as the fire chief/CEO you occasionally need wise counsel from individuals who have held similar positions and walked similar roads. These individuals are not part of your chain of command but rather fellow leaders you have met along your journey. They have lived out the weight of responsibility that goes along with being in the top spot and are willing to share their insights and experiences. As a fire service leader, you need to develop a group of individuals whom you can count on for wise counsel.

BUILDING YOUR NETWORK OF WISE COUNSELORS

It is done differently by different people, but the key aspect in building a network of wise counselors is forming relationships. For me, much of my network came from the fact that I love to teach. I have for a long time been employed part time as a field staff instructor for the University of Illinois Fire Service Institute (IFSI). This world-class training institution uses field staff instructors to provide classes on a variety of topics to fire service agencies. The institute has a tremendous facility in Champaign (IL) located on the campus of the University of Illinois. Numerous props are constructed at this location that allow instructors to teach anything from live burn evolutions to a variety of other fire service disciplines.

But many classes are delivered remotely at local departments. These remote classes are my favorite! Instructors travel to the department, meet with the leadership team, and then are given latitude to tweak the training to make it most applicable for the agencies participating in the course. This tweaking requires that instructors talk and actively listen to department leadership in order to assess their needs. This talking and listening is what I love. Some of my dearest friendships have come from relationships developed through teaching.

Another area where I have found great connectedness is through an international fire service networking think tank that I am blessed to be part of. The group meets twice a year to discuss issues facing the fire service and to think globally about ways to address some of our most challenging issues, including those issues that members see as potential future concerns. Due to the influence of many members, the theories and concepts developed during these meetings often flow into national standards and best practice policies.

What is unique about this group is the fact that many of the members are well past their years of stretching hose and making hallways, but their mental faculties and thinking are as sharp and as cutting edge as you will find anywhere in the world. This group is one of the ways they stay connected to the fire service, and through these interactions their years of experience and lessons learned are imparted to the next generation of fire service leaders.

Other members attending these meetings are still in the trenches, striving daily to make the fire service better, stronger, more resilient, and better able to meet the changing needs and demands of the global fire service. Often this latter group of members come to the meetings beat up, exhausted, and in desperate need of a battery recharge. Because of the relationships that exist between the members of this group as well as the fact that many of the members literally wrote the book on a number of fire service leadership and operational issues, the knowledge and wisdom of this collective assembly is electric and amazing to watch. The relationships I have with these leaders has become a huge part of both my continuing professional development as well as my personal support system. Their encouragement drives me to keep doing the job even when it gets hard, not to mention the fact that their experience and willingness to share has become an endless source of wise counsel for me. One longtime member of this group who recently passed used to say that these meetings serve as the "kitchen table for fire chiefs." Wise words!

WITHSTANDING THE WEIGHT OF THE FIFTH BUGLE

Relationships with those who can provide you with wise counsel and encouragement are such an integral part of the success of a fire chief/CEO that without them the likelihood of failure is high. After watching many fire service leaders try and fail, I have come to believe that it is impossible to go it completely alone (fig. 10–3).

Figure 10–3. It's impossible to succeed without wise counsel, says the author, shown riding in the chief's buggy in the Fourth of July Parade.

You need the collective wisdom that comes through relationships with others who care about you and are willing to spend their time helping you succeed. Sometimes these individuals provide advice, friendship, a listening ear, and a simple word of encouragement. Regardless of what it is ... you absolutely need it if you are going to be successful in carrying the weight of the fifth bugle.

REFERENCES

Powell, C. (1995). *My American Journey*. Random House Publishing Group.

Truman, H. S. (1952). *National War College* [Speech]. Washington, DC.

QUESTIONS FOR FURTHER RESEARCH, THOUGHT, AND DISCUSSION

1. Describe a situation in which you experienced the emotional weight and loneliness of command.

    a. Describe the decision-making process you used to lead through the situation.

    b. What learning occurred through this situation?

2. From your personal network, list three individuals whom you regularly utilize to provide wise counsel in helping you lead your organization.

    a. _____
    b. _____
    c. _____

3. What traits are exhibited by the individuals listed in question 2 that make you go to them for wise counsel?

4. How do you make yourself available to be wise counsel for someone else?

5. When asked to fill the role of wise counsel for someone else, what steps do you take to ensure that you are providing sound advice?

EPILOGUE

Although I dreamed of writing a book for decades, the final push came from my friend David Woodward and his wife, Elizabeth. David and Elizabeth are the owners of One Warrior, LLC. Their firm produces and manages the conference Revolutionary Fire Tactics at the Lake. The conference is held annually at the beautiful Lake of the Ozarks, Missouri (fig. e–1).

In planning for the inaugural conference David asked me if I would be willing to develop a program specifically designed for the fire chief/aspiring fire chief. We both recognized that there are lots of training classes that focus on strategies, tactics, and techniques for handling incidents, as well as a myriad of classes on leading personnel, but most never address how you actually manage a fire service organization. We did not want to just address theories in this class but rather place great emphasis on the actual skills needed to serve as a fire service CEO.

By design the conference has a very hands-on focus, so I felt compelled to build a hands-on class for the fire chief/CEO. I wanted to get the students doing work so that when they went back home, they would have developed some new management skills rather than just listening to me talk. I also wanted to get them thinking with a more CEO-focused mindset where they would look deeply at some of the challenging issues facing today's fire service. I promised David that I would give it my best attempt.

The initial presentation lasted for eight hours and looked at things like strategic planning, succession planning, employee recruitment, hiring, why employee performance evaluations do not work, and what might be a better

Figure e–1. David and Elizabeth Woodward, owners of One Warrior, LLC (photo courtesy of One Warrior, LLC).

option. Through facilitated workgroups the students conducted a strategic planning SWOT Analysis, wrote strategic planning goals and objectives, built questions for employee interviews, and developed options for replacement of the annual performance evaluation model used by many in the class. We also talked about the importance of messaging and the need to see around corners to avoid leading our organizations into a disaster based on a CEO misstep. We hit on the hard reality that many government leaders (e.g., managers, elected officials, other department heads) do not see the fire service through the same rose-colored glasses that we wear. They do not see us as heroes riding shiny and well-polished horses. They see us as expensive, overstaffed, and overequipped, with a huge lobbying force that does nothing but drive unfunded mandates that take up critical financial resources that could be better spent elsewhere. We openly talked about this challenge and worked to come up with ways to address these misconceptions. The students worked hard, and at the end of the day we were all exhausted!

I left hoping that I had challenged them and had facilitated a level of learning, but I also did not expect that I would be back to do this type of training again. I believed that most fire officers would not come to a conference held in a wonderful resort area to learn this kind of stuff. Yet when the course evaluations came in, the reviews were really pretty great. What shocked me the most was that the students had not only made nice comments but also provided a list of additional topics they wanted me to cover in the future. Topics they asked for included financial management, labor/management relations, how to write policies, how to conduct internal investigations, and how to do employee discipline. To my total surprise, David kept saying, "Nobody is teaching this stuff." So, another course, and back to the conference for round two and again for round three.

I do think there are a few training institutions that are trying to teach some of these skills, including the National Fire Academy. The challenge is that no reference book exists for the local fire chief to use in figuring out how to do this type of work. Many fire chiefs/CEOs are juggling so many different hats that they cannot find the time to attend classes or a conference on these topics. Some also face the fact that their organizations cannot financially find the dollars required to send them to this type of training. Hence the writing of this book...

My goal is that this book not only teaches management skills but that it will be used as a how-to instruction manual that provides examples that can be tweaked to fit a department's/district's unique needs. I also hope that my stories serve as case studies that encourage the reader to think deeply about their own situations to find solutions.

I would love to have covered numerous further issues based on requests driven by students, but you can only put so much in a single book. More topics include how to retain talented employees, managing the promotional process, the fire service's role in economic development, data-driven decision-making, the need for formalized education, and so on. I will save those topics for the next book.

As I close, I want to leave you with three thoughts:

1. We are extremely blessed to get to be part of the fire service. Do not ever take that fact for granted.
2. Being the fire chief/CEO is risky and difficult, but it is absolutely worth the risk. I have loved wearing five bugles. Getting to set the standard on how my organizations have touched and impacted people's lives is an incredible experience. I would do it again in a second!
3. Never forget—it is about the people. The people you impact and the lives you touch are what really matter, and they are your true legacy as a leader.

God's Blessings!

INDEX

A

ability to work in United States 88
abrasiveness 94
accountability 136
acknowledgment forms 238–239
acting company officers (ACOs) 81, 162, 166
acting up 81, 126, 133
action steps 4–5, 34, 44
actions unbecoming 216
ADA (Americans with Disabilities Act) 110, 237
adaptability 83
administrative assistant 8, 172
administrative leave 188, 211, 221, 280
administrative skills xix, 122. *See also* organizational leadership skills
    areas of department 131
    development of 133
    policies 228, 229
    soft skills 83
administrative staff 11, 131
admission of guilt 196
adverse employment decision 206, 214, 232
AFG (Assistance to Firefighters Grant) 55
age of candidate 89, 107
AHJ (authority having jurisdiction) 111
alcohol testing 193
alcohol usage 109
AmazonSmile® 61
ambulances
    percentage multiplier 72
    staffing 14
    strategic planning 23–24
    transport billing 58–59
    treat/no transports 58

Amburn, Barry 9
Americans with Disabilities Act (ADA) 110, 237
Ames, Jeanine 99
Anderson, Larry L. 21
annual budget 39, 46, 76. *See also* budgeting
annual performance review 165. *See also* performance evaluations
    generations 154
    real-time feedback vs. 157, 160–161
    review period 150
apparatus. *See also* ambulances
    command vehicles 68, 72
    engine 68
    fire trucks 38
    heavy rescue truck 47
    ladder trucks 51, 68
    pickup truck 38
    preowned and donation 47
    pumpers 71, 72
    replacement 23, 36, 67
    sinking fund 68–69
    tower ladders 68, 70, 72
    trailers 72
    water tenders 38
application fee 86
applications
    allowed and prohibited information 87–89, 104, 107
    for grants 56
    written 86–87
arbitration 214, 217, 239
arson 109
assertiveness 110
asset life expectancy 67, 69
Assistance to Firefighters Grant (AFG) 55

assistant chief 81, 186, 214
attention to detail 83
attorneys. *See* legal counsel
authoritarianism 95
authority, circumventing 263
authority having jurisdiction (AHJ) 111
Ayers, Glen 242, 245

B
baby boomers 156–157, 181
back cover 36
background investigations 82, 90, 105–107
backstep position 246
Ballestra, Nicholas 99
battalion chief 81, 130–132, 172
beliefs 13. *See also* values
Bennett, Nathan 133
bequests 56
bias
    internal investigations 189
    messaging 13
    performance evaluations 151–152, 181
    rater 152, 181
billing 25, 57–58
Bill of Rights 201
blood testing 193
blueprints 5, 26
blue shirts 82, 100
board of directors
    hiring and 7
    majority vote 266
    relationship with 257, 261, 264–266
boogeymen mentality 267
books 226
branding 56
Brunacini, Alan 123, 223–224, 269
Bryant, Jeff 55
Bstow.com 61–62
budget cycle 51, 64
budgeting 39, 46. *See also* financial management; revenue
    capital items 66
    expenses 49
    fundraising 55

    line items 49, 64
    long-term savings 50
    maintenance 52
    planning 21
    salaries 131
    tracking 64–66
building permits 63
bullet points vs. narrative statements 30
business administration xviii
business license inspections 63
buy-in 29, 148

C
Candidate Physical Ability Test (CPAT) 61, 93
candidate pool 81, 86
capital 257–259
capital asset planning 66–67
capital budget items 66
capital equipment replacement program spreadsheet 36, 68, 78
Cappelli, Peter 160
career development
    personal 142–143, 285
    planning 118, 135–136
    stages of 126
career fire departments 2, 118
    community expectations 50
    them vs. us mentality 267
carrot-and-stick program 149, 165
case files 212–214
case law 233
case studies 5–6
    management 259–261
    performance evaluations 161–162
    360° evaluations 175–179
CBA (collective bargaining agreement) 130, 165
cellar nozzle 240–242
CEO (chief executive officer) xviii. *See also* fire chief/CEO
certificate of occupancy 63
certifications 88, 97
chain of command xvii, 216, 263
change management 39–40, 176, 178, 262

Index

changing conditions 14, 42, 153
charitable organizations 57, 62
chief executive officer (CEO) xviii. *See also* fire chief/CEO
circumventing authority 263
citizenship 89
city council 22
city managers 163
    managing money and 50, 73
    relationship with 187, 257, 261–264
    strategic planning with 22
    succession planning and 124
Cleveland Board of Education v. Loudermill 215
clinical interviews 110
coaching
    career development 134
    day-to-day 159
    discipline 208, 210, 220
    follow-up email 212
    performance evaluations 153, 158
Code Black 225
Code Green 225
code of conduct 239. *See also* policies
Code Red 225
Code Yellow 225
collaboration 12
collective bargaining agreement (CBA) 268
    grievance procedure 217, 235
    Loudermill hearings 215
    part-time members 130
    pay raises 165
Collins, Jim 22
Collins, Robert 281
combination fire departments 8, 118
comfort zone 138
command staff 121, 261. *See also* executive leadership
command vehicles 68, 72
common law test 85
communication 269
    candidate skills 83
    challenging conversations 263
    complaints 192

    demands 11
    evaluation 174, 177
    face-to-face 155
    guiding documents 226
    job performance 158–159
    sharing plans 132
    written 228
community education program 57
community livability 84
community members. *See* constituents
community risk assessment 21
community rooms 60–61
company pride 126
comparison line chart 174
    leadership 175
    leading change 176
compassion 97
compensatory time accrual 3, 16
complaints 192–193, 269
    form 193
    property damage 195
    receiving 190, 192, 219
    rudeness 208
    self-initiated 192
    tardiness 208
    theft 206
    trespassing 192
composure 110
Compton, Dennis 267
comradery 191
concerts 60
conditional job offers 103
conduct, rules of 205
conferences 60, 289–291
confidentiality 161, 173
confined space entries 60
connectedness 11
    between departments 132
    networking 284
    with personnel 40, 266–267
constituents
    expectations 11, 17
    SWOT Analysis with 33
    willingness to pay 50
construction permit fees 63

contact information 88, 193, 195
contract policy development firms 239
convalescent transports 58
cooperativeness 168
coordinators 131–132
coping skills 110
corporate sponsorships 56
cost recovery program 24, 59
council-manager government 261, 264
cover page 34
COVID-19 66
coworkers 172
CPAT (Candidate Physical Ability Test) 61, 93
CPR classes 61, 217
Craig, Rodney S. 265
Crandall, Brian 81
creativity 83, 176, 178
credentials 188, 211, 282
credit check 89, 105
criminal activity 192, 203
criminal history 89, 105
criticality 137, 143
crosslay hose load 243
Culbertson, Tori 157
customer relations 169, 174, 195. *See also* public relations

D

Davis, Brent 9
decision-making
    feedback 174
    guiding documents 225, 246
    internal investigations 187, 206
    weight of command 287
de-escalation 195
degree, master's 134. *See also* education
delegation 126, 246
demotion 211, 221
dependability 168
deployment 244
deputy fire chief 8
Developing the Leaders Around You 129
development 26
diabetic patients 58

Dillon's Rule 52, 75
direct supervision 95
discipline 205–212, 291. *See also* termination
    carrying out 208–209
    challenge 280
    changing behavior 218
    cumulative 208
    demotion 211
    disciplinary action form 213
    documentation 150, 163, 212, 221
    employee coaching 210
    internal investigation 188
    policies 207–208
    progressive 207, 209–212
    relief of duty 211, 221
    SOGs 231
    state and federal regulation 220, 237
    suspension 210, 217, 220
    unpaid suspension 209, 217
    verbal warning 209, 210, 220
    written warning 209, 210, 220
disclaimer 232
discriminatory hiring practices 87, 107. *See also* hiring
discriminatory policies 233. *See also* policies
dismissal 150
dispatch recordings 193
disposition of the case 198–201
disqualifying backgrounds 105, 116
division chief 172
documentation
    acknowledgment forms 238
    case files 213–215
    check-in meetings 162, 262
    discipline 150, 212, 221
    paper trail 163–165, 212
    performance evaluations 163–165
    preliminary investigation 193
documented verbal warning 209, 210, 220
dress code 98
drill training 239, 245
driver/engineer 246
driver's license 87

driving record 105
driving under the influence 62, 186, 209
drug testing 193
drug usage 109, 111
dual role battalion chiefs 130–132
Dyer, Smokey 271

E

EAP (employee assistance program) 237
Eby, Wes 256
e-commerce 61
economic downturn 73
education 3
   background 88, 97
   community program 57, 217
   graduate 134
   paramedic school 100, 118
   technology degrees 134
   for upper levels 122
EEO (equal employment
     opportunity) 237
efficiency 245
egos 268–269
80/20 Rule for Hiring New Team
     Members 96–97
elected officials
   communication with 266
   election cycles 255, 262
   expectations 11, 266
   hiring and 83
   level of service 50
   managing money 73
   relationship with 258, 274
   succession planning and 124
eligibility lists 103, 137
emergency management 228, 229
emergency medical services (EMS) 92
   battalion chief 131
   certification classes 61
   event standby 60
   hiring for 108
   policies 228, 229
Emergency Operations Center
     (EOC) 229
emergency preparedness 118
emergency response 227
emotional triggers 94
employee assistance program (EAP) 237
employee handbook 2, 225, 232–239
   disclaimer 232
   individual topic format 236, 249
   running category format 235, 249
   templates 234–235
   writing 236
employee performance. *See
     also* performance evaluations
   competencies 168–169
   expectations 207, 232, 239
   misconduct 192
   poor 150
   succession planning 123
   tardiness 208
   termination 214
   traits 114
   valuing good 80–81, 149
employees. *See also* post-offer testing;
     testing; turnover
   benefits 17, 124, 237
   development 126
   probationary period 8, 167
   problem 80, 170, 206, 259–260
   remote working 156
   rights 201, 233
   turnover 115, 155, 160, 206
employment law 234, 240
EMS. *See* emergency medical services
     (EMS)
engagement, rules of 271
engine company operations 227, 243–244
engines 68
envelope financial management 48–50.
     See also financial management
EOC (Emergency Operations
     Center) 229
equal employment opportunity
     (EEO) 237
equipment
   donations 47
   preowned 47
   protective 65, 67, 227

equipment (*continued*)
   replacement 67, 68
   surplus 47
escrow accounts 68
essential functions 88
estate planning 56
evidence 193, 198
exams. *See* testing, written
Excel 69
executive leadership 6, 259. *See also* command staff; leadership
exoneration 188
expenditures 52
   budget tracking 64
   capital 66
   future 66–68
   nonoptional 49
   unforseen 65
experiential learning 122, 134
Explorer Post Program 117

F

Facebook. *See also* social media
   candidate 107
   orientation via 91
   recruitment via 83
facilitator 30
Fair Labor Standards Act (FLSA) 2, 16, 124, 237
families 124, 135, 156
Family Medical Leave Act (FMLA) 237
farm team 128–130, 141
Federal Emergency Management Agency (FEMA) 55
feedback
   for employees 150, 153
   from mentors 136
   real-time 154–155, 158
   for supervisors 176–179
   SWOT Analysis 30
   tool 174
"fee for service" model 24, 57
fees
   application 86
   construction permits 63
   fines 54, 62
   infrastructure rental 61
   patient care 58
   plan reviews 63
   self-pay 58
   sign permits 63
   special event standby 60
   training 61
   water/sewer tap 63
FEMA (Federal Emergency Management Agency) 55
field training officers (FTOs) 100, 217, 284
   performance evaluation 170
   program 95
Fifth Amendment 201
fight-or-flight response 109
financial management xix, 45. *See also* budgeting; revenue
   envelope 48
   guiding principles 48
   strategic planning and 51–52
   training 291
fines 54, 62
fingerprints 107–108
fire academies 61, 95, 100
fire administration 118, 121
fire chief/CEO xviii–xix
   direct report 163, 261–264, 275
   discipline 205, 209
   elected officials and 51
   experience 256
   figuring it out 4
   hats to wear 11, 291
   hiring 80
   internal investigations role 190–191
   loneliness of command 282–283, 287
   managing money 45
   outside 128, 259
   public pressure 190, 216, 255
   references 108
   relationships xiv, 254
   selecting officers 7–8
   succession planning 118
   tenure 255–256

training xiii, 289–291
your why 10
Fire Chief Magazine 242
fire code violations 54
Fire Command 124, 223–224
fire departments. *See also* career fire departments; volunteer fire departments
   combination 8, 118
Firefighter Hearts United 47
Firefighter Hiring Act 102
firefighters
   career 117, 147, 278
   female 91, 92
   full-time 2, 124, 129
   male 91
   minority 91, 92
   multigenerational 80, 83, 102, 117
   paid 83–84
   part-time 128–129
   reserve 277
   typical model 123
   volunteer 2, 80, 84–85, 148
firefighter's association 57, 61
fireground
   size-up 225
   support 227
fire marshal 118
fire protection technology 118
fire safety inspections 63
fire science 134
fire stations 23
fire suppression 59
fire truck 38
First Aid classes 61, 217
First Call Rule 270–271
fitness for duty 111
501c3 entity 57, 61
flexibility 155
Flint, James 254
FLSA. *See* Fair Labor Standards Act (FLSA)
FMLA (Family Medical Leave Act) 237
Forbes 133

forms
   acknowledgment 226, 238–239
   complaint 194
   disciplinary action 213
   internal investigation report to the chief 202
   investigative case activity log 200
   investigative case control sheet 199
   joint supervisor evaluation 168–169
   self-evaluation 171
   standard operating guideline template 230
   sworn statement 197
fortitude 97
Fourteenth Amendment 201
fraud 58, 105
Freedom of Information Act 63
FTOs. *See* field training officers (FTOs)
full-service emergency response agencies 92
full-time firefighters. *See* career fire departments; combination fire departments; firefighters
fundraisers 21, 47, 55
funeral memorial gifts 56
future, planning for the
   leadership skills 133, 142, 176, 178
   manager and 263
   strategic planning 26
   SWOT Analysis 28
   visioning 13, 15
future value 69

G

gap rating 175
Gardner v. Broderick 203
Garrity Rights 202–203
Garrity v. New Jersey 202
gender bias 152
gender of candidate 89
general statutes 54
generations
   baby boomers 156–157, 181
   evaluation process and 153–156
   Generation X 156–157, 181

generations (*continued*)
   Generation Z 153, 155–156, 181
   millennials 154–155
goal/objective chart 34–35
goals
   career 135, 142
   evaluation 169, 176
   financial management 51
   organizational 163
   performance evaluations 148
   personal 159, 163
   SWOT Analysis 33–35, 36
Good to Great 22
goodwill 257–259
government
   council-manager 261, 264
   finance 46
   public bodies 50–51, 121, 290
   strong-mayor 261, 265
   taxation 52–53
grants 55–56
grievances
   employee handbook 235
   example 217
   First Call Rule 270
   internal investigations 201
   steps of process 221
grit 23–24
group discussions 29
guiding documents 225–227. *See also* policies
   employee handbook 232–237
   job performance requirements 240–246
   standard operating guidelines 227–232
   struggle with 246
guilt, admission of 196

H
Haigh, Beth 9
hair length 233
Haithcock, Scott 9
handbook. *See* employee handbook
Handy, V. Keith 8
Harvard Business Review 23, 98, 101
Hatzold, Tom 6

hazard mitigation 59
hazardous materials response 59, 62
health and sanitation inspections 63
health care administration 134
health care plan 2
hearings 214
heavy rescue truck 47
Heifetz, Ronald 39
Heim, Joseph M. 20, 25, 199
hiring 79
   application 86–89
   candidate potential 82
   discriminatory practices 87, 107
   eligibility lists 103
   interview 95–100
   legacy 112–113, 125
   orientation 90–92
   from outside the department 127–128, 139, 259
   part-time personnel 130
   physical ability test 92–94
   post-offer testing 103–112, 115
   preference points 102–103, 130
   race of candidate 89, 107
   references 82, 88, 108–109
   religion of candidate 89
   sexual orientation of candidate 89, 107
   team 82–83, 101–102
   types of testing 104
Home Rule 52–53, 75
honesty 259
hose lead-out 244–245, 247–248
hose load 243
hostility 40
human relations xviii, 92, 94, 134
human resources
   hiring team 83, 99, 102
   internal investigations 189–190
   performance evaluations 153

I
IAFF. *See* International Association of Fire Fighters (IAFF)
IAPs (incident action plans) 229
ICS (incident command system) 229

identifying needs 22–23, 26, 42
idiosyncratic rater effect 152, 158, 170, 181
IDLH (immediately dangerous to health and safety) 38
Illinois Fireman's Disciplinary Act 216
Illinois Fire Service Institute (IFSI) 1, 12, 254, 284
immediately dangerous to health and safety (IDLH) 38
incident action plans (IAPs) 229
incident command system (ICS) 229
inconsideration 95
individual topic format 235–236, 249
inflationary multiplier 67, 69
infrastructure
   development 26
   rental fees 61
injuries 92, 185, 231
inspections
   business license 63
   services for 63, 228, 229
insubordination 198, 203
insurance 3, 58–59
integrity 174
intellectual abilities 110
interest income 72
internal affairs 189–190
internal conflict 31, 268
internal investigations 189–205, 291
   case files 212–214
   complaint 192–193
   disposition of the case 198–201
   employee misconduct 192
   employee rights 201
   example 216, 280
   fire chief/CEO role 190–191
   organizational culture 191
   preliminary investigation 193–196
   questions 197
   report to chief 201
   struggles 189–190
International Association of Fire Chiefs 93, 268
International Association of Fire Fighters (IAFF) 93, 128, 130, 270
interpersonal skills 83, 110
interrogations 203, 213, 219
interview 82
   clinical 110
   dress code 98
   face-to-face 97
   investigative 204
   law enforcement 107–109
   process 96–100
   questions 98, 115, 290
   remote 97
   stress 98–99
   team 101–102
   techniques 95–96
invalid assist calls 58
investigating officer 193
investigations 189. *See also* internal investigations
   case activity log 198, 200
   case control sheet 198
   preliminary 193–196
   questioning 189
investing in team members 129
issue date/revision date 230, 234

J
jealousy 151
job history 88
job offer
   conditions 103
   enhancements 84
job performance requirements (JPRs) 3, 225
   example 241–242, 251–252
   time and motion studies 240–245
joint supervisor evaluations 166–169
journal notes 150, 161, 165, 212
JPRs. *See* job performance requirements (JPRs)
Junior Firefighter Program 117

K
Kelly Days 129

key stakeholders 27
kitchen table advice 281
Knight, Rebecca 98, 101

L

labor board 204
labor law 87, 107. *See also* legal counsel
ladder trucks 51, 68
Landry, Tom 159
language proficiency 88
lapse in judgment 186, 206, 216
Lasky, Rick xiv
latchkey kids 156
law degree 134
law enforcement
   interview 107–109
   investigation 189–190, 220
   relationship with 217
leadership. *See also* command staff
   challenges 279–280
   engaged 264–265
   evaluation 174
   executive 6, 259
   long tenure 253
   poor decisions 127
   potential 82, 95, 153
   questions 175
   senior roles 120
   skill development xiii, xviii–xix, 133
   storytelling in 6, 127
   strategic planning and 26, 27
   them vs. us mentality 267
   union 268
   unwillingness 138
Leadership on the Line 39
learning
   experiential 122, 134
   plan 136
lease-purchase agreements 69
leave
   administrative 188, 211, 221, 280
   paid administrative 188, 211, 216, 221
   sick 129, 186
   unpaid administrative 211, 221

   victim's economic security and safety 237
legal action 212
legal counsel 134
   case files 214, 220
   employee discipline 150–151, 163
   employment law firms 234, 240
   labor attorneys 87, 189, 234
   policy review 236
   state and federal standards 237
   tax attorneys 54
LEIE (List of Excluded Individuals/Entities) 105
level of service 50–51
levy-based taxes 53, 59
Lexipol 239
liability 134, 226
licenses 88
lieutenant 8, 81, 130
life expectancy of asset 67, 69
life experience 82, 97, 284
line items 49–50, 64
line-of-duty deaths 111
Linsky, Marty 39
list-making 25
List of Excluded Individuals/Entities (LEIE) 105
litigation 188, 231, 234
live burn training 23
living document 22
loneliness of command xvii, 282–283, 287
Loudermill hearings 214–215, 222

M

mailbox 194
maintenance 52, 65, 67
malicious act 62
malicious intent 206
management. *See also* city managers; financial management
   change 39–40
   of different generations 154
   new broom sweeps clean 259–261
   performance evaluations 151
   policy 239

project 21
reactive vs. progressive 14
time 83, 243
turnover 262
manuals 226
marital status 89
marketing 62, 84
marriage 2, 9
master's degree 134. *See also* education
mathematical reasoning 94
Maxwell, John 129
mayor 187, 264
measure of success 34, 44
mechanical reasoning 94, 100
media 11
   employee misconduct and 187, 188, 216
   public relations 61
   recruitment and 84
Medicaid 58
medical billing companies 58
medical evaluations 111
medical history 89
medical necessity of need 58
medical response protocols 3, 25, 92
Medicare 58
meetings
   check-in 159, 162–163, 262
   discipline 205
   group 147, 279–280
   Loudermill 215
   one-on-one 167
   SWOT Analysis 29, 33
memorial gifts 56
memory testing 94
mentoring 122, 159
   career development 136
   fire chief 281
   Generation Z 155
   leadership 134
   millennials 154
   succession planning 126–127
mergers 63, 127
merit rating system 153
Merritt, Michael 8

messaging 13–14, 18, 36
metro departments 118, 147
microdonation programs 61–62
micromanagement 156
Miles, Stephen A. 133
milestones 161
military experience 88, 102–103, 261
military rating system 152–153
military reserve 89
millennials 154–155, 181
minimization 109
Minute-Man hose load 244
misconduct 192
mission
   assessing 17
   employees 80, 267, 269
   level of service 51
   in the strategic plan 26, 36
   visioning 4
Moore, Heather 34
motivation 97, 99, 148
multigenerational firefighters 80, 83, 102, 117
municipal code 62
My American Journey 282

N

narrative statements vs. bullet points 30
National Fire Academy 6, 134, 291
National Fire Academy's Personal Analysis and Development Plan 135–136
National Fire Protection Association (NFPA) standards 70, 83, 130
National Fire Protection Association's—Fire Service Needs Assessment 55
National Labor Relations Board (NLRB) v. J. Weingarten, Inc. 204, 220
National Public Employer Labor Relations Association 234
neglect of duty 192
neighborhood canvas 107
Ness, Eliot 7
networking 284
network of counselors 255
neutral reference policies 108

new broom management theory 259–261
new candidate orientation 90–92
NFPA 1582: Comprehensive Occupational Medical Program for Fire Departments 111
NFPA standards. *See* National Fire Protection Association (NFPA) standards
NLRB (National Labor Relations Board) 204
no-call/no-show 186
nonemergency transports 58
non-fire-focused responses 92
nonoptional expenses 49

O

objectives 34, 44
observation 94
Occupational Safety and Health Administration (OSHA) 59
office hours 11, 264
officers
    coordinators 131
    development programs 61
    discipline 208, 209
    hiring 81
    investigating 193
    performance evaluations 166
    selecting 7–8
    succession planning 120
    support of 148
onboarding process 106
1000 Ways to Recruit Top Talent 95
one-time revenue
    fundraisers 55
    grants 55–56
    memorial gifts 56–57
    sponsorships 56
One Warrior, LLC 289
on-site standbys 60
open-mindedness 178, 179
operations
    battalion chief 131
    budget 65
    policies 228, 229

opportunities 27, 32, 43
ordinances 54
organizational culture
    candidate fit 90
    change 256, 260
    discipline 205
    disruption 211
    grit 24
    internal investigations and 191
    internal struggle 31
    leadership 265–266
    long-term success 83, 137
    policies 237
    professionalism 85, 177
    succession planning 125
    values 96, 128, 177
organizational instability 40, 145
organizational leadership skills 121, 122, 257. *See also* administrative skills
orientation, new candidate 90–92
OSHA (Occupational Safety and Health Administration) 59
outside facilitator 30–31
outside investigator 201, 214, 220
overtime 3, 16
    administrative duties 131
    budgeting for 64–65, 131, 185
    planned vs. unplanned 66

P

paid administrative leave 188, 211, 216, 221
paid-on-call departments 124
paid personnel 83–84
paper trail 163–165, 212
paramedics 19
paramedic school 100, 118
part-time firefighters 128–129. *See also* firefighters
passion 24, 97, 136, 174
passivity 95
patient care fees 58
patient care protocols 3, 25, 92
patient care reports (PCRs) 208
pay raises 149, 165–166. *See also* salaries

payroll 64
PCRs (patient care reports) 208
pencil pushers 123
The Pennsylvania Fireman 47
pensions 3
people development 160. *See also* career development
percentage multiplier 70–72
percentage-spent calculation 64
perception of performance 170, 175
performance evaluations 147. *See also* annual performance review
    building a case against an employee 150–151
    case study 161–162, 175–178
    definition 149
    discipline 206, 212
    history of 152–153
    joint supervisor evaluations 166–169
    quarterly reviews 38
    questions 158, 161–162, 182
    real-time 157–158
    self-evaluation 167, 170
    semiannual reviews 38
    teams 157
    360° evaluation 170–175
    truth and bias 151–152
performance rating system 153
performance time standard 240–243, 252
perjury 196
perseverance 24
personal accounts 196. *See also* sworn statements
personal growth 129, 256
personal references 108
personal responsibility 191, 209
photographs
    documenting complaints 193, 195
    in the strategic plan 38
physical ability test 92–94
physical strength 92
pickup truck 38
PIO (public information officer) 229
planning 169. *See also* strategic planning
plan review fees 63

police chief 22
police officers 102, 190
police powers 62
policies 291
    categorizing 228–229, 229, 236–237
    development firms 239
    discipline 207–208
    discriminatory 233
    distribution of 239
    guiding documents 226
    historical copies 231, 234, 249
    legal validity 233–234
    neutral reference 108
    review 230, 250, 251
    violations 192, 207, 217
political capital 257–259, 273
politics 262
    affiliation of candidate 107
    connections 259
    within organization 31
    pressure on fire chief xiv, 120, 124, 255, 266
    southern vs. northern communities 3
polygraph investigations 109
postincident analysis 158, 160
post-offer testing
    background checks 105–107
    fingerprinting 107–108
    fitness for duty medical evaluations 111
    polygraph investigations 109–110
    psychological testing 110–111
    reference checks 108–109
    typical conditions 103–104
Powell, Colin 254, 282
power of arrest 62
power shift 129
predisciplinary hearing 214
preference points 102–103, 115, 130
pregnancy status 89
preliminary investigation 193–196
preowned equipment 47
presenters 91
pride 97, 126
private insurance 58

private investigator firms 190, 220
private/public partnership 57, 61
private sector experience 261
private service providers 23
probationary period 8, 167
problem employees 80, 170, 206, 259–260
problem-solving 83, 254
productivity 168
professional growth 129
progressive discipline 207, 209–212
project management 21
promotions
   assessment center 8
   hiring and 81
   list 137
   succession planning and 126, 132
   from within 129, 139
property damage 194, 198
property taxes 52–53. *See also* taxes
Property Tax Extension Limitation
      Act 53
protective equipment 65, 67, 227
psychological testing 110–111
public administration 261
public education training 217
public information officer (PIO) 229
public policy 134
public relations 11, 276. *See also* customer
      relations
   community rooms and 61
   employee misconduct and 187, 211
   recruitment and 84
public safety messages 56
public safety psychological
      testing 110–111
public works director 22
pumpers 71, 72

Q

quarterly reviews 38–39. *See*
      also performance evaluations
questions
   candidate orientation 90–92
   career goals 135–136
   check-in meetings 159
   discipline 206, 219–222
   Garrity rights 203
   hiring 114–116
   internal investigations 197, 219–222
   interviews 95–96
   leadership 175
   leading change 176–177
   Loudermill hearings 215
   managing money 75–78
   performance evaluations 158, 161–162,
      181–183
   policy development 249–252
   polygraph 109
   relationships 273–276
   strategic planning 42–44
   succession planning 133, 141–145
   360° evaluation 173–174
   time and motion studies 243–244
   visioning 12–13, 16–18
   wise counsel 287–288
   written test 94–95

R

race of candidate 89, 107
radio recordings 195, 196
rallies 60
rate-based taxes 53
rater bias 152, 181
reading comprehension 94
real-time evaluation system 157
recruitment
   paid personnel 83–84
   part-time personnel 130
   problems 105, 114
   street-level work vs. administrative
      skills 122
   volunteer personnel 84–85
references 82, 88, 108–109
referendums 53
reflection time 12–13, 30, 135, 150
regulations 54
rehab 231–233
relationships 291. *See also* public relations
   with the boss 261–264
   case study 259–261

with elected board 264–266, 275
with law enforcement 217
with other organizations 255, 283
with personnel 266–270, 276
political capital xiv, 257–259
with problem employees 260
with team 257
trusted leader 254, 273
with the union 186, 216, 258, 268–269
wise counsel 285
reliability 168
relief of duty 211, 221
religion of candidate 89
remote working 156
rental venues 60–61
replacement of equipment 67, 68
report cards 149
reserves 65
resistance to change 39–40
response district 243
responsibilities
discipline 209
fire chief 257, 283
goals 34, 44
increased 162
planning 24
policies 232
responsiveness to authority 110
retirement 137, 143, 156
revenue 49. *See also* budgeting; financial management; one-time revenue; sustained revenue; taxes
alternative 54–55
from ambulances 14
one-time 55–57
planning 21, 23, 76
sustained 57–62
taxes 52–53
review process of documents 250
employee handbooks 233
SOGs 230
revision date 229, 230
Revolutionary Fire Tactics at the Lake 289

risk
community 21
overtime 131
strategic plans 40
visioning 10
road map 22, 150
Roberson, Steven 117
roll call training 239
rookie school training program 100, 239
rules of conduct 205
rules of engagement 271
running category format 234–235, 249

S

SAFER (Staffing for Adequate Fire and Emergency Response) 55
safety awareness 169, 269
salaries
battalion chiefs 131
budgeting 65
competitive 154, 269
demotions 211
division chief 131
leadership roles 124
pay raises 149, 165–166
sales tax 54
salvage services 59
savings account 38, 65
SCBA (self-contained breathing apparatus) 38, 67, 72
scheduling 124
SEIU (Service Employees International Union) 128, 130, 270
self-contained breathing apparatus (SCBA) 38, 67, 72
self-evaluation tool 167, 170
self-focus 95
self-incrimination 203
self-initiated complaints 192
self-pay fees 58
self-reflection 135, 172, 178
semiannual reviews 38–39. *See also* performance evaluations
senior staff 91

Service Employees International Union (SEIU) 128, 130, 270
sex offender registry 105, 108
sexual orientation of candidate 89, 107
shared values 96, 125. *See also* organizational culture
shift coverage 66
shift vacancies 66
shotgun testing 86
SHRM (Society for Human Resource Management) 153
sick leave 129, 186
side-by-side rating system 167
sign permit fees 63
silent generation 153
sinking funds 38, 68–72, 78
situational judgment 94
size-up 225
Skills and Maintenance Manual 227, 240
skill sets 80
   evaluating 167
   job performance requirements 227, 240
   of leaders 133, 142
   soft skills 83
small tools 65
Smith, Forest 56
social life 135
social media 154, 188. *See also* Facebook
   background check 107
   recruitment 83
   Twitter 83
   YouTube 83
Social Security Administration 85
social security number 90
Society for Human Resource Management (SHRM) 153
SOCs (Station Observation Checklists) 242
soft skills 83. *See also* administrative skills
Spare Change 61
spatial orientation 94
special event standby fees 60
sponsorships 56
sporting events 60
squads 72

staffing
   lack of 11, 185
   paramedic 14, 19
   succession planning 136
Staffing for Adequate Fire and Emergency Response (SAFER) 55
stakeholders 27
stamina 92
standardized format 230
standard operating guidelines (SOGs) 3, 225, 227–232
   attachments 232
   categorizing 228–229, 229
   definitions 231–232
   discipline 231
   guidelines & information 232
   issue date/revision date 230–231
   purpose 231–232
   review of 230
   scope 231
   template 230
   writing 228
standard operating procedures 223, 226. *See also* standard operating guidelines (SOGs)
standing orders 225
state tax laws 52–53, 54. *See also* legal counsel
Station Observation Checklists (SOCs) 242
status reviews 38–39
statute of limitation laws 249
stopping the clock 187–188, 216
Stop the Bleed classes 61, 217
storytelling 5–6, 127
strategic planning 290
   budgeting 39
   celebrating success 40–41
   example 19–20, 26–27
   execution and 24–25
   financial management and 51–52
   professional look 34–36
   questions 42–44
   status reviews 38–39
   SWOT Analysis 27–33

time frame 34
street-level work 122–123, 134
strengths 27, 31, 43
strong-mayor government 261, 265
structural PPE 67
subordinates 172
succession planning
   at all levels of organization 121, 127
   challenges 119–122, 141
   delegating duties 126, 246
   example 126
   farm team 128–130, 141
   hiring and 81, 125
   priority 137, 143
   tracking members 136–138
   worksheet 138, 144
successor plan 40
Sullivan, John 95
supervisors
   bias 152, 181
   discipline 208, 209
   goals 163
   joint 166–167
   performance evaluations 158, 166–167
   relations 95
   360° evaluations 170–175
support system 9
surplus equipment 47
SurveyMonkey 173
suspensions 210, 217, 220. *See also* discipline
sustained revenue
   ambulance billing 57–58
   community rooms 60–61
   cost recovery programs 59
   fees for service 59
   infrastructure rental fees 61
   microdonation programs 61–62
   special event standby fees 60
   team of record contracts 59–60
   training fees 61
sworn statements 196, 203, 219. *See also* witnesses
SWOT Analysis 27–33, 36, 290
   with community 33
   conducting 29–30
   example 31–32, 43–44
   labor/management relationship 274–275
   team 28, 36, 42
   writing goals 33–35
sympathetic nervous system 109

T

tactical plan 244, 247–248
tailboard riding 70
tardiness 208. *See also* employee performance
target completion date 34, 44
TargetSolutions 238
tattoos 233
Tavis, Anna 160
taxes 2, 48
   caps 53, 75
   elected officials and 262
   fees for service vs. 59
   levy-based 53, 59
   property 52–53
   rate 53
   rate-based 53
   revenue from 52–53
   sales 54
   state laws 52–53, 54
   willingness to pay 51
team members 7–9
   as recruitment tools 85, 91
   SWOT Analysis 28, 36
   tracking 136–137
   union and 268
team of record contracts 59–60
teamwork 246
   generations 154
   hiring 83, 97, 99
   performance review 161
technical rescue 59
technology degrees 134. *See also* education
telecommunication companies 61
templates
   employee handbook 234–236
   SOGs 230

termination 188, 212–214, 221
  discipline policy 207, 211
  example 216
  weight of command 282
test development 94
testing
  custom 95
  job performance requirements 240
  physical ability 92–94
  psychological 110–111
  shotgun 86
  written 94–95
theft 109, 206
Thorpe, Barbara 8
Thorpe, Ron 8
threats 27, 32, 43
360° evaluation 170–175, 182
  case study 175–178
  confidentiality 173
  data analysis 174, 175–176
  implementation 172
  question development 173
time and motion studies 240–245
time demands 10–11, 125
time frame, plan 34
timeline, case files 214
time management 83, 243
tolerance 95
tower ladders 68, 70, 72
traditionalists 153
traditions 155
trailers 72
training 12
  battalion chief 131
  candidates 96
  discipline 206, 208
  fees 61
  financial management 46
  fire administration 122
  fire chief/CEO xiii, 289–291
  instructors 284
  job performance requirements 240–242
  online 239
  policies 228, 229
  program 132–134
  public education 217
  roll call 239
  succession planning 132–134
  timed 242
treat/no transports 58
trespassing 192
trust 46–47, 254, 256
trusted leader relationship 254
turn-in lease program 69
turnover
  employee 115, 155, 160, 206
  manager 262
Twitter 83. *See also* social media

U

underlying medical problems 111
unforeseen expenditures 65
uniforms 65, 91
union
  contracts 84, 130
  dues 217
  employee handbook 235
  expectations 269
  hiring 102
  negotiations 130, 268
  relationship with 186, 216, 258, 268–269
  representation 204, 215, 216
  succession planning 125
unpaid administrative leave 211, 221
unpaid suspension 209, 217. *See also* discipline
untouchables 7
U.S. Fire Administration 223
U.S. Inspector General—List of Excluded Individuals/Entities (LEIE) 105

V

vacancies
  acting up 81
  shift coverage 66, 128
  succession planning 137
  volunteer departments 280
values 90, 96, 135. *See also* organizational culture

Varone, Curt 203, 218
vehicle. *See also* apparatus
   extrication 59
   fires 59
venues 60–61
verbal warning 209, 210, 220. *See also* discipline
VESSA (victim's economic security and safety leave) 237
Veterans of Foreign Wars (VFW) Post 265
veterans' preference 102
victim's economic security and safety leave (VESSA) 237
visioning
   definition 4–5
   evaluating 175, 177
   messaging 13–14
   organizational 12, 18
   questions 16–18
   steps 5
   struggle with 9–10
   your why 10, 18
volunteer fire departments 2, 19, 253, 277
   administrative leave 188
   employee handbook 235
   employees 79–80, 84–86, 124
   level of service 50
   motivation 84–85
   them vs. us mentality 267
volunteer work 88
voters 50, 53

W

waivers 92, 93
warning 210, 220
water main 26–27
water/sewer tap on fees 63
water tender 38
weaknesses 27, 31, 43
websites 83

weight of command xvii–xix, 216, 279–283, 285–286
Weingarten Rights 204–205, 220
whistle blower 263
white shirt potential 82, 100
why, evaluating your 10
willingness to pay 50, 75
wise counsel 187, 283–285, 287–288
witnesses. *See also* sworn statements
   direct report and 264
   investigation 198, 280
   Loudermill hearing 215
   preliminary investigation 193, 195
Woods, Robert 9
Woodward, David 289
Woodward, Elizabeth 289
work
   avoidance 95, 110, 138
   environment 13, 176, 177
   ethic 83, 139, 149
   groups 29, 172
   history 88, 110
   retreat 12
   schedule 124, 156
workaholics 156
workforce 154–157, 181
work-life balance 135, 154, 156
workplace practices 237
written application 86–87
written decision 215
written documents 246. *See also* guiding documents
written report 201, 262, 273
written statement 196, 215
written testing 94–95
written warning 209, 210, 220. *See also* discipline

Y

Yancey, Rick 3
YouTube 83. *See also* social media

ABOUT THE AUTHOR

Craig A. Haigh began his fire service career in 1983 as a volunteer in his hometown of Hampton (IL). He served as Hampton's volunteer fire chief from 1991 to 1995. During that time, he developed and implemented their EMT-Intermediate/Advanced EMT program. In 1988 he was hired full time as a firefighter with the city of Rock Island (IL) Fire Department. He ultimately was appointed as their first EMS coordinator and as such developed and implemented the department's paramedic program. He left Hampton and Rock Island in 1995 to become the first paid fire chief of the King (NC) Volunteer Fire Department. There he was tasked with transitioning the all-volunteer department into a combination agency (volunteer and paid employees). While in King he served as the primary architect responsible for merging the independent volunteer department into the city of King, making it a municipal government agency. He returned to Illinois in 2002 to serve as fire chief for the Village of Hanover Park. He was the 2012 Illinois Career Fire Chief of the Year.

Chief Haigh also works as a field staff instructor with the University of Illinois Fire Service Institute (IFSI), where he has been since 1995. In his work at the University of Illinois he is a regular partner with IFSI Research and Skidmore College First Responder Health and Safety Laboratory. He is the recipient of the 2019 International Association of Fire Chiefs Alan Brunacini Fire Service Executive Safety Award, given for his work in developing operational practices based on scientific research in an effort to reduce firefighter deaths and injuries due to cardiovascular/medical events.

In addition to his duties as Hanover Park's fire chief, Chief Haigh has served as the interim village manager and was also tasked with the creation of the Inspectional Services Division. This new division merged fire prevention, the building department, and the health department into a single agency under the control of the fire department. He also created, implemented, and has managed the village's strategic planning program.

Chief Haigh became the emergency management director of Hanover Park, Illinois, in January 2020 in addition to his job as fire chief.

Chief Haigh retired in 2021 from full-time service as a firefighter and now serves as an independent consultant focusing on management and organizational leadership. He presents and speaks internationally. In retirement, he has started volunteering as a firefighter/paramedic with the East Dubuque (IL) Fire Department.

He has a bachelor of science degree in fire and safety engineering and a master of science degree in executive fire service leadership. He is a graduate of the United States National Fire Academy Executive Fire Officer Program and is a nationally certified paramedic, accredited chief fire officer, and member of the Institute of Fire Engineers. He is married to his wife, Beth, and has two adult children, Melody and Lee.